Nisbet Ebenezer

**The Science of the Day and Genesis**

Nisbet Ebenezer

**The Science of the Day and Genesis**

ISBN/EAN: 9783337138493

Printed in Europe, USA, Canada, Australia, Japan

Cover: Foto ©berggeist007 / pixelio.de

More available books at **www.hansebooks.com**

# THE SCIENCE OF THE DAY

### AND

# GENESIS.

#### BY

## E. NISBET, D.D.,

AUTHOR OF " RESURRECTION OF THE BODY: DOES THE BIBLE TEACH
IT?" ETC.

Prove all things. — PAUL.

Science must consist of precise knowledge. — HUXLEY.

ROCHESTER, N. Y.:
C. VENTON PATTERSON & CO.,
1886.

INSCRIBED

TO

# MY SISTER, E. F. NISBET.

" We 've clamb the hill thegither."

(5)

# CONTENTS.

# PREFACE.

THIS treatise has a unique aim: it deals with all points of contact between science and the Bible history of creation. A brief but thorough *résumé* of the latest teachings of science in this entire field is presented and reviewed. A large amount of matter, elsewhere only to be gathered by extensive reading, is here found. The volume will be of service to persons who would be abreast with current thought, pastors, and Sunday-school workers.

Interpretations of Bible are not always Bible; hypotheses in science are not always science. This volume is intended to render human interpretation more precisely the Divine thought, and aid in the distinguishing of Nature's assured utterances from the chatterings of pseudo-science.

Human interpretations of the Bible and science, pseudo-science and the Bible, may conflict, — science and the Bible never. It is no less false than unjust to both scientists and biblicists to denominate scientists "sceptics, opponents of the Bible, infidels." It belies many scientists, and gives the impression that biblicists are clinging desperately to something which thinking has outgrown; that there is a fight going on here between a blind, credulous, traditionary faith, and the light of facts. Nothing more false: the leaders in inductive science in America to-day are thorough Bible believers, — Profs. J. D. Dana, Asa Gray, J. W. Dawson. Biblicists are not to ignore this, and throw the weight of these men and their facts into the enemy's camp. Says President Chadbourne, "The difficulty in reconciling the geologic with the Mosaic record is found by students of the Bible who know nothing practically of geology, and by students of geology who are ignorant of the Bible, or hate its plain requirements, so that they wish to discredit the book for their own peace of mind. Devout men, eminent both as practical geologists and Bible students, like Hugh Miller, Dawson, Dana, do not seem to find any real difficulty in the case. If there are any two living men who are better able to give an opinion in this matter than Profs. Dana and Dawson, I, for one, should be glad to go far to take lessons from them." That this volume will help to confirm this position of President Chadbourne, is believed by

THE AUTHOR.

LEAVENWORTH, KANSAS, June 1, 1881.

(9)

# THE SCIENCE OF THE DAY AND GENESIS.

---

## CHAPTER I.

### WHENCE THE EARTH?

MODERN science has offered us solutions of "Whence the earth?" Descartes, two centuries ago, suggested the hypothesis of vortices. In the Cartesian system, a vortex is a collection of matter forming an ether or fluid endowed with a rapid rotary motion around an axis. From such vortices Descartes constructs the universe.

A second hypothesis, the nebular, was first proposed by Kant, developed somewhat by Herschel, thoroughly systematized by Laplace (obiit 1827). He supposes that the space of our solar system was once filled by matter of high temperature, and rarefied much more than our most rarefied gases. This was the primary nebula. This nebula moved on its axis, cooled by throwing off heat into space, and as it cooled, condensed. The circling mass, becoming flattened at the poles and bulged out at the equator, assumed gradually the form of a disk. The attraction of cohesion of the matter at the circumference of the disk was finally overcome by the centrifugal force; this outer matter was separated from the central condensing mass, and became a revolving ring. Ring after ring was thus formed, constituting by their condensation our now planets. The rings earliest liberated were of matter less condensed than those thrown off later from the ever-consolidating mass; and so we find the planets Uranus and Neptune, far distant from the solar central mass, have the specific gravity of cork, while the other planets increase in specific gravity as they near the sun. The pale rings of Saturn to-day may give some conception of the disintegration of the primary nebula.

After the ring containing the matter of our earth broke and fell into itself, the globe thus constituted had an immense dilatation, embracing our moon. The matter finally constituting our earth, revolving and condensing, became a globe of melted lava. Then, after indefinite centuries, a scoria formed here and there on the surface of the cooling mass. By and by, the scattered scoriæ unite. A crust is formed enclosing a sea of fire. From time to time, by contraction of the ever-cooling core, the crust breaks, and lava belches forth; or the crust is corrugated, and incipient mountains are formed. The surface of the earth ever cooling more and more, finally the vapors, gases, etc., of the atmosphere begin to fall; seas, lakes, rivers are formed; by their attritions soils appear, then vegetation, then creature life of lowest forms. The interval during which the terrestrial crust would be lowered from 2,000° to 200° has been estimated by Helmholtz at three and one half millions of centuries.

Such is the nebular theory of Laplace. He claimed that certain spots of light in the heavens, of duller lustre than the stars, were examples of matter even now existing in a nebulous state; but when later, Lord Rosse's immense telescope was applied to these supposed nebulæ, Laplace's entire brilliant fabric tottered. It was found that some of the spots of light supposed to be nebulous matter were resolvable into distinct stars. It was now claimed that it only needed a telescope of sufficient power to resolve all the so-called nebulæ into stars. The hypothesis fell into disrepute. But the recent revelations of the spectroscope declare that there are in the heavens genuine nebulæ; and the nebular origin of our solar system has again come into favor. It is now generally maintained by scientists.

Previous to the time of Laplace, Sir Isaac Newton had said, "The admirable arrangement of the solar system cannot but be the work of an intelligent and most powerful Being." Laplace claimed that Newton in this statement "had deviated from the method of true philosophy"; and it has been thought that Laplace, in propounding the nebular hypothesis, had an atheistic purpose in view, seeking by his theory to indicate how our solar system might have originated without the aid of "an intelligent and powerful Being." While granting to Laplace the eternity of matter, we may yet ask him, Whence the laws impressed on that matter, causing to spring forth our solar system in such beauty, harmony of movements, adaptations, stability? Does not all this speak of more than a mere eternal sovereign force? Does not all this speak of a preconceiving, prearranging, powerful, sovereign Intelligence?"

Further, grant Descartes his vortices, Laplace his nebula: these had motion, they tell us, — a rotary motion. That motion could not have been from eternity; had it been from eternity, the vortices must have remained vortices, the nebula a nebula, to eternity; eternal uniformity of motion past must have ever retained matter in eternal uniformity of condition. That these vortices, that that nebula could have taken new shape in time, some force not impressing them from eternity must have wrought on them in a new beginning; there must have existed some force outside of themselves. Was not that new outside force, touching them in time, what Genesis calls "God"?

So the heat of the nebula could not be from eternity, else it would have remained uniform to eternity. Was not the new power outside of the nebula, throwing in heat upon it in time, what Genesis calls "God"?

Grant to Laplace the nebular origin of the

earth, this nowise conflicts with Genesis. Says
the nebular hypothesis, " Our globe was once a
formless, unfurnished, chaotic mass, brooded over
by darkness." Says Genesis, " The earth was
once emptiness and voidness; darkness was the
swaddling band thereof." Later (so-called)
science is but an echo of earlier Genesis.

Genesis not only claims that a force, power,
outside of matter has given our world and its
present furnishing their shaping, but it also
claims that that power is an Intelligence, a personal Being, — God; and science to-day, in its
most authentic expounders, reiterates the Mosaic
assertion. Will, Intelligence, a Person, gave this
universe birth. Says Agassiz (speaking of the
life furnishing of the earth), " The combination
in time and space of these thoughtful conceptions,
exhibits not only thought, — it shows premedita-
tion, power, wisdom, greatness, prescience, omnis-
cience, providence; in one word, these facts in
their natural connection proclaim aloud the one
God, whom man may know, adore, and love : and
natural history must in good time become the
analysis of the thoughts of the Creator of the

universe, as manifested in the animal and vegetable kingdoms."

Dr. W. B. Carpenter, one of the foremost scientists of Great Britain, regards "Nature, or the material universe," as "the embodiment of the divine thought," and "the scientific study of nature" as "the endeavor to discover and apprehend that thought" (to have "thought the thoughts of God" was the privilege most highly esteemed by Kepler); and Carpenter quotes approvingly Mr. Martineau: "What, indeed, have we found [in scientific research] by moving out along radii into the infinite? That the whole is woven together in one sublime tissue of intellectual relations, geometric and physical, — the realized original, of which all our science is but the partial copy; that science is the crowning product and supreme expression of human reason. . . . Unless, therefore, it takes more mental faculty to construe a universe than to cause it, to read the book of Nature than to write it, we must more than ever [in the late revelations of science] look upon its divine face as the living appeal of thought to thought." Carpenter, as

Agassiz, regards the universe as "a revelation of the mind and will of Deity," and that that Deity is a person.

So Alfred Wallace, the co-propounder with Darwin of the present prevailing phase of the evolution hypothesis, says, "It does not seem improbable that all force may be will force, and thus that the whole universe is not merely dependent on, but actually is the will of higher intelligences, or of One Supreme Intelligence. It has been often said that the true poet is a seer; and in the noble verse of an American poetess we find expressed what may prove to be the highest fact of science, the noblest truth of philosophy : —

> ' God of the granite and the rose !
>    Lord of the sparrow and the bee !
> The mighty tide of being flows
>    Through countless channels, Lord, from thee ;
> It leaps to life in grass and flowers,
>    Through every grade of being runs,
> While from creation's radiant towers
>    Its glory flames in stars and suns.' "

Thus do science and Genesis harmoniously declare : This earth and its furnishing have arisen from the will of an Intelligence, from God.

Further, in the expression of Genesis, "*God* created the heavens and the earth," there is declared the *unity* of the power pervading all space and all time; and precisely this is the utterance of science to-day, whether we look at the uniformity and harmony of the operations of the universe in the present, — the evident concatenated development of an original one plan in the readings of the geological rock record of the past, — or the recent doctrine of the correlation of forces, that the sum of force in the universe, potential and actual, is always one and the same, a unit. "The men who did most to prepare the way for this doctrine, the correlation of forces, such as Newton, Davy, Oersted, Herschel, and Faraday, all delighted to see God in his works; and the philosopher who was the main agent in discovering it, Dr. Mayer, has a mind filled with the presence of God, and looks on force as the expression of the divine power." Says Dr. Carpenter, "The culminating point of man's intellectual interpretation of nature may be said to be his recognition of the *unity* of the power, of which her phenomena are the diversified manifestations. Towards this point all scientific inquiry now tends."

Science to-day, in one of its most recent fundamental advances, the correlation of forces, is just beginning to grasp in its full significance, in its sublime height and all-encompassing breadth, the grand idea enunciated in the first verse of the Bible cosmogomy, "One power pervades all things." And we have recently offered us by Saigey a volume whose object is to demonstrate the "Unity of Physical Forces," in which vital activity itself is made simply transformed motion. Of the "atom and motion" he would construct the universe.

Says Prof. Tyndall, "I have noticed, during years of self-observation, that it is not in hours of clearness and vigor that this doctrine [materialistic atheism] commends itself to my mind; that in the presence of stronger and healthier thought it ever dissolves and disappears, as offering no solution of the mystery in which we dwell and of which we form a part." A confession, this, that the most advanced scientific thinking of the day, which unifies all the forces producing the varied phenomena of the universe, cannot, in its "stronger and healthier" hours, persuade itself that the all-pervasive Power, the ultimate Producer, is non-intellectual force; demands something higher, finds rest only in the acceptance of the Genesis utterance, "An Intelligence, God, created the heavens and the earth."

Says Dr. Carpenter, "In the admirable words of the great master, Sir John Herschel, 'In the only case in which we are admitted any personal knowledge of the origin of force, we find it connected (possibly by intermediate links untraceable by our faculties, yet indisputably connected) with volition, and by inevitable consequence with motive, with intellect, and with all those attributes of mind in which personality consists.' As a physiologist," continues Carpenter, "I most fully recognize the fact that the physical force exerted by the body of the man is not generated *de novo* by his will, but is derived from the oxidation of the constituents of his food. But holding it as equally certain, because the fact is capable of verification by every one as often as he chooses to make the experiment, that in the performance of every volitional movement, that physical force is put in action, directed and controlled by the individual personality or *ego*, I deem it just as absurd and illogical to affirm that there is no place for a God in nature, originating, directing, and controlling its forces by his will, as it would be to assert that there is no place in man's body for his conscious mind." "Sun, fire-mist, molecules," says Carpenter, "but what is back of the molecules? A personality." Genesis names that personality "God."

# CHAPTER II.

## THE AIM OF THE BIBLE.

To interpret the Bible correctly in its connections with science, it is imperative that we understand and keep in mind the mission of the Bible, and its method of dealing with man in his primitive scientific status.

The mission of the Bible is distinctively and absolutely spiritual. The Bible is written simply to save man from sin and its consequences. Teaching physical science is thus wholly foreign to the Bible. It comes to man, not to interfere with his ideas of physical science, but, meeting him on his own peculiar plane of knowledge physical, adapts itself in infinite condescension to his child-like, inadequate ideas of the mysterious universe of matter in which he finds himself, and takes ho'd of the hand of the wandering child with simply one thought; to lead him back to

the Father he seeks; leaving the correction of
his ideas of the physical universe, and of all other
human sciences, to the growing light of the child's
developing reason, and the unfoldings of time.

Accepting this as the correct view of the case,
we are not to expect that the Bible, given to man
in his infancy, is to address that man as if he
were acquainted with the Copernican system of
the universe, had weighed with Torricelli the
firmament, and explored with Lyell the rocks.
These fields are left gymnasia for the play of
man's God-bestowed intellect.   If the divine wisdom, come to teach man moral truth, find him
in his infancy in physical knowledge, holding
the idea that the expanse of the earth's atmosphere is a sol.d crystalline vault in which the
stars are set, and in which are " windows of
heaven," if he thinks the earth stands still and
the sun moves, — the language is adapted to such
views.

It has, in opposition to this view, been urged
that a divine teacher, without descending to the
office of teaching science, might yet have kept
his own language free from all collusion with

human error. In reply to this, De Quincey well says : "Meantime, if a man sets himself steadily to contemplate the consequences which must inevitably have followed any deviation from the customary erroneous phraseology of the people, he will see the utter impossibility that a teacher (pleading a heavenly mission) could allow himself to deviate one hair's-breadth (and why should he wish to deviate?) from the ordinary language of the times. To have uttered one syllable, for instance, that implied motion in the earth, would have issued into the following ruins : *First,* it would have tainted the teacher with the reputation of lunacy ; *secondly,* it would have placed him in this inextricable dilemma : on the one hand, to answer the questions prompted by his own perplexing language would have opened upon him, of necessity, one stage after another of scientific cross-examination, until his spiritual mission would have been forcibly swallowed up in the mission of natural philosopher ; but on the other hand, to pause resolutely at one stage of this public examination, and to refuse all further advance, would be, in the popular opinion, to

retreat as a baffled disputant from insane para-
doxes which it had not been found possible to
support. One step taken in that direction was
fatal, whether the great envoy retreated from
his own words, to leave behind the impression
that he **was** defeated as a rush speculator, or
stood to his words, and thus fatally entangled
himself in the inexhaustible succession of expla-
nations and justifications. In either event the
spiritual mission was at an end; it would have
perished in shouts of derision, from which there
could have been no retreat and no retrievance
of character. The greatest astronomers to-day,
rather than seem ostentatiously learned, will
stoop to the popular phrase of the sun's rising
and setting; but God, for a purpose commen-
surate with man's eternal welfare, is by some
critics thought incapable of the same petty
abstinence."

Accept this view as to the method of the Bible's
dealing with man, in reference to the facts of
nature, there is at once removed all scientific
objection to the inspiration of the Bible from its
*incidental* expressions touching nature, incorrect

in fact; *e. g.*, the constitution of the firmament, "windows in heaven," immovability of the
earth, etc.   And if we find the grand volunteer
utterances of its opening page — the creation of
all by the One whose book the Bible claims to be,
and creation in a certain order in the creatures
and time — confirmed by later science, we have
here evidence that a wisdom higher than the
human wisdom of that early day wrote.   The
Book will thus on its first page declare its divine
Author, and give a reason for a respectful hear-
ing.

# CHAPTER III.

## THE ANTIQUITY OF THE EARTH.

GEOLOGY, as an inductive science, had its origin as late as 1807, in the formation of "The London Geological Society." In some matters it is only yet seeking its ποῦ στῶ; in others it has reached solid standing. It has disabused the mind of the idea that the earth as it is now constituted was spoken instantaneously into existence by immediate divine fiat. That not divine fiat immediately and instantaneously, but natural agencies mediately and by long process have built up the earth's crust in its present form, is proven by what we find in that crust, collated with what we see taking place continually about us. The geologist sees on the sea-beach, delta, lake shore, or bottom to-day, layer after layer of sand gently laid down; he sees now forming calcareous rocks embedding implements of man's art and man himself, the fossils of the future; he sees wood buried becoming lignite in its process of transformation into coal; he sees the bones of existing species of animals buried in sediment; he sees volcanoes thrusting forth their melted rock; he sees corals forming their islands, reefs, atolls, and shells agglutinating; he sees the workings of chemical affinity and. voltaic action. He now passes from the surface of the earth down into its depths; he turns over leaf after leaf of the stone book. On each leaf he sees distinctly traced the impress and seal of the very agents he finds to day in action at their incessantly modifying work on the crust of the earth all round about him. The rocks of the earth's crust are now to him a new revelation. As he turns over their stone leaves, he recognizes the earth's own handwriting; he reads there its autobiography, — a writing that can be no forgery. Says the geologist, " I find verified in the records of nature what I find written in the records of the Bible, ' I change not!' As to-day the wind and the wave and the cloud, the sunshine, heat, cold, winter's snow, summer's rain, autumnal sleet, vernal shower, belching volcano, the little

unseen, silent-working coral insect, are all ministers in the hand of the one universal intelligent
Worker, modifying the crust of the earth and for-
warding his world-idea; so down deep in the
earth's crust I find traces of these same ministers
accomplishing mediately, and by process of duration incalculable, the behests, in earth-crust modification, of the one sovereign Intelligence, and
from out the rocky deeps I hear a voice I have
heard before, — 'I am the same yesterday, to-
day, and forever.' "

But if natural agencies have fashioned the
earth's crust as we now find it, a duration of
indefinite length is needed: needed for the deposition of the twenty miles of stratified rocks in
that crust; needed for the deposition of the immense accumulations of shells in these rocks,
nearly one seventh of their entire bulk; needed
for the rise, life, and dying out (slow processes)
of nearly fifty different worlds (Agassiz) of creatures, which have successively peopled our globe,
leaving their traces in its crust; needed for the
growth and deposition of the immense vegetable
accumulations in the coal measures, in Nova Sco-

tia, with their interstratified soils, nearly three miles in thickness (Lyell). Six million years have been claimed for the coal measures alone.

The heavens above us declare that the physical universe is created on a grand scale as to time. Light travels nearly 200,000 miles a second. It takes light from nine to twelve years to reach us from the nearest fixed star. Herschel discovered stars whose light must travel 9,000 years before reaching our world, and nebulæ whose light would only reach us after 3,000,000 years. Indication here of the grandeur of the time scale of our universe; one part becoming aware of the creation of the other part only after the expiration of 3,000,000 years.

This voice from the stars may lessen our astonishment at the voice from the rocks demanding for our earth an antiquity of millions of years.

The point in time of the creation of our world is not given us in the Bible, nor any data by which it may be approximately determined. The only Biblical statement is, "In the beginning God created the heavens and the earth."

## CHAPTER IV.

### "DAY" IN GENESIS I.

LONG before the rise of geology, "day" in Genesis i. was judged difficult to interpret. Josephus declared "day" metaphorical; Origen thought it an indefinite period; Augustine declares it not only difficult to understand, but even to conceive what sort of day is here meant. In more modern times, "day" has been regarded a figurative expression for an indefinite period; others regard it the ordinary day; others still think it was understood by the writer as an ordinary day, but stood in the Divine mind as symbol of a vast period.

Any interpretation of "day," to be accepted, must harmonize with both of God's volumes, — Nature, the Bible.

(*a.*) Those who hold that the "day" of the narrative of creation is an ordinary day maintain that an indefinitely long period intervenes between the creation spoken of in the first verse, and the creations recounted in the verses immediately following, — these latter verses recounting simply the creations which took place a few thousand years ago at the introduction of man. The gap thus left, it is claimed, between the first verse and the verses following gives ample time for all the worlds of geology.

This view assumes that in twenty-four hours our continents rose from an unbroken ocean and took their present form; that in forty-eight hours all the now existing species of animals came into being. Science, on the contrary, declares that such processes demand immense duration. This interpretation is now generally rejected. Dr. Conant, in rejecting it, declares the assumption of a long lapse of time between the creative act of the first verse and the creative acts of the verses which follow to be wholly unwarranted by anything in the sacred writer's statement; and claims that the extension of the creative work through six successive periods, of whatever duration, can be

explained only by the fact that the work was not
accomplished by a sudden exertion of supernatural
power, but "by the operation of those secondary
causes which the structure of the earth proves to
have been active in its formation, requiring ages
for their accomplishment."

(*b.*) The figurative interpretation, making
"day" an indefinitely long duration, agrees with
science ; but the narrative, by the exact limita-
tion "evening" and "morning," impresses us with
the idea that the writer conceived of "day" as an
ordinary day. The same impresses us at the insti-
tution of the Sabbath (Ex. xx. 10, 11).

(*c.*) The symbolical interpretation of "day" —
"day" *understood by the writer* as an ordinary day,
but standing in the Divine mind as a symbol of a
higher duration — solves all difficulties.

Some of the foremost Biblical scholars maintain
that the creation history was communicated to man
in successive visions, — tableaux. The six tableaux
of creation rose before the eyes of the seer, impressed him as six successive periods of work, —
as six ordinary days, and their rise and fading
away as morning and evening. But in the Divine

mind these six tableaux were symbols of periods
of past working of indefinite length.

Revelation of God's works past holds the same
relation to the human and the Divine thought as
revelation of God's works future. The same
method of interpretation is applicable to both.
One is prophecy teaching backward, the other
prophecy teaching forward. We find the "day,"
"week," "year," of prophecy forward stood sometimes in the Divine mind — the event infallibly
interpreting — as symbols of higher periods (Dan.
ix. 24-27, and xii. 11, 12; Ez. iv. 6). So if the
unfoldings of God's works in the past by physical
science teach that in the symbol "day" there lay
in the Divine mind an outlook and conception
infinitely more grand than the human language
would indicate, or the human mind then was fitted
to receive, we are to accept these unfoldings of
science as giving us God's own interpretation of
the miniature symbol "day." As the child-man
advances in physical knowledge, the height and
grandeur of the full content of the miniature
symbol open up to him, just as the unfoldings
of historical events lead man to the height and

grandeur of the Divine thought in their miniature symbols.

And here are perfectly harmonized the ordinary day conception of the writer of the narrative, and the indefinitely long periods required by science. Each of the six tableaux of the creative week impressed the seer as an ordinary day, while in the Divine mind each tableau was symbol of an indefinitely long period.   And the facts of the rock record correspond precisely with the Mosaic tableaux in the *kind* of creations, and in the *number* and *order* of the creation periods.  Traces of the first, second, and third day's work — light, firmament, the heavenly bodies — must be only incidental in the rocks; full traces of only the third, fifth, and sixth days' work — vegetation, creeping creatures, beasts, and man — can we expect the rocks to give us.  Full traces of these latter three days' work we find in the rocks, and in the precise order given us in Genesis.  The geologic scale divides itself into three grand parts : Palæozoic, Secondary, and Tertiary.  The Palæozoic — corresponding to Genesis' third day — was emphatically the plant period, "herbs yielding seed after their kind." In no other age was such vegetation; this is the period of vast vegetable accumulations constituting our coal. On the fifth creative day appear the sea monsters, creeping creatures, birds. Corresponding with this, the Secondary period of geology abounded above all other periods in enormous monsters of the deep, creeping creatures, and birds of wonderful size; and coincident with the beasts, cattle, man of the sixth creative day, the Tertiary is specially the epoch of great beasts, of cattle, and the only epoch in which traces of man are found.

The great divisions of the geologic scale thus correspond in number and in kind of creatures and order of creation with the three symbolic tableaux of Genesis — the third, fifth, and sixth days; and the "morning" and "evening" of the Biblical narrative find their antitypes in the gradual introduction and gradual fading out of the peculiar existences of each of the great periods.

## CHAPTER V.

### CREATION OF SUN, MOON, STARS.

THE writer describes things not as they absolutely *are*, but as they would have *appeared* to an unskilled observer on the surface of our earth, — as they *appeared* to the eye of the seer in the vision tableaux. In this principle we have the explanation of the narrator's putting the creation of the "light bearers" on the fourth day, although light existed from the first day.

When our globe was simply a glowing fire-ball of melted lava, the heat thrust out from it, into the surrounding space, vapor, gases, and other materials, enshrouding the earth in a dense cloud: "The cloud was the garment thereof, and thick darkness the swaddling band." That cloud garment must have remained wrapping the earth during vast, indefinite periods. During all these periods the "light" only of the "light bearers" would reach the earth; no disk yet of sun, moon, star. As these disks first became visible through the enshrouding cloud envelope of the earth, they would *appear* to an observer on the earth just then to have been created, and so the seer wrote: the time of the *appearance* to him in the fourth tableau of their disks, he makes the time of their creation.

The rock record accords with this. The vegetation of the third day — that of our coal measures — was a rank and flowerless vegetation, well adapted to a warm, steaming atmosphere, muffled in cloud. Immediately after this, vegetation changes, as also animal life; by both changes there is indicated a change towards dryness and intenser light in the atmosphere, as well as a change in the constitution itself of the atmosphere, rendering it suitable for the respiration of the new animals now appearing.

This is the explanation given of the existence of "light" during the first three days, by St. Basil, St. Cæsarius, and Origen, long before geology had birth: the "light bearers" existed during the first three days, but their disks were not visible through the earth's cloud mantle. "Who," exclaims Origen, "that has sense can think that the first, second, and third days were without sun, moon, or stars!"

A condition of atmosphere analogous to that I claim for our earth down to the close of the coal period is found amid the Andes of South America to-day. "A thick mist during a particular season obscures," says Humboldt ("Cosmos"), "the firmament, for a period of *many months*. Not a planet, not the most brilliant stars of the southern hemisphere, — Canopus, the Southern Cross, nor the feet of Centaur, — are visible. It is frequently almost impossible to discover the position of the moon. If by chance the outlines of the sun's disk be visible during the day, it appears devoid of rays. According to what modern geology has taught us to conjecture regarding the ancient history of our atmosphere," continues Humboldt, "its primitive condition as to its mixture and density must have been unfavorable for the transmission of light."

To claim that the earth remained in a heated, steaming condition down to the close of the coal

period may seem to demand an exorbitant time
for its cooling; but speaking of the cooling process to-day, Prof. Dana says ("Geology," 683) :
"At present very little of the interior heat of the
earth reaches the surface.  According to Poisson,
the amount is only one seventeenth of a degree
Fahrenheit; and to reduce this amount one
half, or to one thirty-fourth, would now require
100,000,000,000 years.    Mr. Hopkins, of England, has stated that, supposing this the only
mode of cooling, it must have required as long
a time as this, 100,000,000,000 years, to have
diminished the earth's temperature the last two
or three degrees of its decrease."  Nor would a
planet covered over for ages with a thick screen
of vapor be a novelty even yet in the universe.
It is doubtful whether astronomers have ever yet
looked on the face of Mercury, — it is at least
very generally held that only his clouds have
been seen.  Even Jupiter, though it is thought
that his mountains have been occasionally detected
raising their peaks through openings in his cloudy
atmosphere, is known chiefly by his dark, thick,
shifting bands.    It is questionable whether a

human eye on the surface of Mercury would ever behold the sun, notwithstanding his near proximity; nor would he be often visible, if at all, from the surface of Jupiter.  Says Dawson, "Jupiter and Saturn are probably still intensely heated, and encompassed with a vaporous 'deep'; untold ages must elapse before they can sustain life like that on the earth."

Prof. Dana maintains almost identically the view here presented ("Geology," 742 *et seq.*).

I prefer this view to that of Whiston, discarded by Newton and Laplace, — that on the fourth day the earth, by a change in its axis, came into a new relation to the sun; also to the view held by some, that the self-illumination of the earth, or diffused illuminated cosmical matter, supplied the light of the first three days (Knapp, Kurtz, Dawson, Winchell).

Prof. Dana claims that the placing of the creation of the sun so long after the creation of light, so accordant with modern science, is strong evidence of the divine origin of the Bible cosmogony; no human mind at that era would have so written.

## CHAPTER VI.

### DEATH AMONG ANIMALS.

THE old heathen poets delighted to sing of a golden age past. So, also, Milton sings of a golden age past, no death yet even for animals, —not until,

> "Her rash hand in evil hour
> Forth reaching to the fruit, she plucked, she eat;
> Earth felt the wound, and nature from her seat,
> Sighing through all her works, gave signs of woe,
> That all was lost. . . .
>                   Discord first,
> Daughter of sin, among the irrational,
> Death introduced through fierce antipathy;
> Beast now with beast 'gan war, and fowl with fowl,
> And fish with fish; to graze the herb all leaving,
> Devoured each other."

The belief that Adam's sin was the origin of discord and death among animals has widely prevailed.

Geology negatives this : —

"Under the ribs of some monstrous fossil reptiles, greatly earlier than Adam, their stomachs are still found, containing the more solid relics of the food on which they had lived. Among these relics of food are the bones and scales of fishes, showing the marks of the teeth of the reptile which devoured them."

A carnivorous animal cannot live on herbage, nor an herbivorous animal on flesh. Are we, then, to suppose that before Adam sinned, lions, tigers, eagles, vultures fed, like oxen, sheep, and sparrows, on herbage, fruits, and seeds? They could not, with their present anatomy. Are we to suppose a miraculous change in their anatomy, at the moment of Adam's sin? We have no evidence of such change, but evidence to the contrary. But had all animals previous to Adam's sin been herbivorous, in the herbage they eat and in the water they drank, unnumbered lives must have perished.

The Bible does not hint that the death of animals is in any way connected with Adam's sin. The Genesis narrative makes no reference to death among animals. But we may fairly infer from it

that death was, from the beginning, by the very
constitution of things, the law of all earthly life,
man included.  That man might escape this law,
God planted a tree in the garden — "the tree of
life" — with power to preserve against death.
This unique provision for the preservation of man
from death forces upon us the inference that
death was the original law of all earthly life,
man included; and to man, only through the
expedient of "the tree of life" was escape possible from the universal law, death, even had he
abode in holiness.

# CHAPTER VII.

## DARWINISM.

THE evolution hypothesis is ancient and of many forms.

The Egyptian sage maintained that our globe was originally a ball of wet clay. The clay drying in the sun, little blisters arose. These, becoming impregnated with some subtile spiritual influence, became the embryos of all future terrestrial organisms. Upon the bursting of the clayey shells, the earth became peopled by creatures of low grade, which in time were developed into the beauty and perfection of the living forms, man included, now inhabiting the earth.

The Greek Epicureans held a similar theory of the rise of terrestrial organisms.

Since the middle of the last century, evolution theories have been rife. The one which obtained

most notice previous to Darwinism was that of
the French naturalist La Marck, propounded at the
beginning of the present century. He claimed
that spontaneous generation from the ocean
produced all organisms, man included. At first
appeared animal life in its lowest forms; and
from that, all we now see has been developed.
He held that effort to act and habit of action are
the great developing force. For instance, some
fowls, by continually making an effort to swim, so
stretched the skin on their feet that they became
web-footed; the heron, on the contrary, disliking
the water, and drawing itself up to keep dry, has
become long-legged; the giraffe, by the habit of
reaching up among the tree limbs for its food,
so stretched its neck that it became permanently
long-necked. Science now declares spontaneous
generation unproven, and La Marck's theory of
development is now laughed at.

The prevailing phase of the evolution theory at
present is Darwinism, propounded about twenty-
five years ago by Charles Darwin of England, in
his "Origin of Species."

He makes "natural selection" the main develop

ing force.   By natural selection Darwin means
simply this : In the struggle for life going ever on
among animals, those individuals having any inju-
rious variation of form perish; those having any
advantageous variation of form survive and prop-
agate their peculiarity; and the advantageously
formed individuals being ever thus selected by
nature from each generation, the animal creation
ever develops into new and higher forms.   Here
we have the origin of species, genera, classes,
man, — one or more primal monads only having
been created by God.

(*a.*)   Does Darwinism conflict with the Bible?

Not in demanding a great age for the earth :
the Bible tells us nothing about the age of the
earth.    Not in its theory of the origin of life on
the earth : Darwin ascribes that directly to God
("Origin of Species," 1st ed. 419).    Not in
claiming that species and genera, man's body included, have arisen by a process of natural selection : the Bible gives us no information as to the
process by which God produced these.   The sep-
aration of the light from the darkness, causing the
alternation of **day and night**; the formation of

the firmament; the elevation of the land from out the waters, — were all effected by slow process of natural forces, and why should not God proceed in the same method in the production of the creatures? Darwin, as the Bible, claims that "All the races of man are descended from a single primitive stock" ("Descent of Man," I. 220). Genesis impresses us that man is a unique being on earth: "God breathed into him the breath of life," he "became a living soul," he was made the "image and likeness" of God. Darwin also declares man unique, the only "moral being" on earth ("Descent of Man," I. 85, and II. 370). Darwin claims that man derived in germ all in him from "a hairy quadruped, furnished with a tail and pointed ears, and an inhabitant of the Old World"; and this creature is descended, "in the dim obscurity of the past, from an aquatic animal provided with branchiæ, with the two sexes united in the same individual." Genesis impresses us that man did not receive his unique, high nature, setting him over all other terrestrial creatures, from any of these subject creatures, but directly at his appearance from God. This matter for the present thrown out, we may come to the examination of the other dogmas of Darwinism without religious prejudice, regarding these dogmas as having simply a scientific interest.

(*b.*)   Does Darwinism present sufficient proof to make it probable, and warrant us to accept it provisionally?   I answer negatively.

Fundamental to Darwinism is transmutation of species.   It is very doubtful whether any vegetable or animal species has ever been observed to put on the characteristics of another species, by either natural or artificial selection.   So Herbert Spencer concedes.   Huxley says, " It is our clear conviction that, as the evidence stands, it is not absolutely proven that a group of animals, having all the characteristics of a species in nature, has ever been originated by selection, whether artificial or natural."   Quatrefages ("Human  Species ") claims this transmutation for a kind of wheat; but says great care in its cultivation must be exercised, or it perishes, — which is a confession that it is not a "good" species.   This non-observation of the transmutation of species is a serious objection to Darwinism;   this  Huxley  confesses,  and  says,

"As the case now stands, this 'little rift within the lute' is not to be disguised nor overlooked."

The rocks give no evidence of transmutation· of species; their voice is against it.

In the earliest geological epochs, indeed, the lower types of organisms predominate; in the later, the higher types. But we do not find, first the low, gelatinous, homogeneous individual, then the partially developed, yet aborted organism, and this, in concatenated progression of forms, passing into the perfectly developed individual of the species, and this species passing by similar process into a new species, genus, family, class. Not a trace of this is found. But just such process must geology exhibit before it becomes auxiliary to the dogma of transmutation of species by natural selection. On the contrary, at the very first appearance of any form of life in the rock record, it is always perfect; often the earlier species are of higher type than the later.

For instance, the marine fucoids — sea-weeds — are the earliest plants known. When later the land vegetation appears, it bears no evidence of being a gradual development of the marine fucoids, but it appears all at once perfect, and a higher development in some cases is found than now: the primeval ferns and club-mosses attained the height of forest trees, now they are pygmies.

Similarly speaks the geological record of animal life. Eozoön Canadense, of the lower Laurentian rocks, is its earliest trace, — mere animated jelly. Now notice, between the first appearance of Eozoön and the lower zone of Silurian life above, there intervene about 100,000 feet of rock, representing, probably, millions of years. Were the Darwinian natural-selection development theory true in nature, should we not expect to find Eozoön, as it mounts up through hundreds of thousands of years towards Silurian life, developing into something higher than mere homogeneous undifferentiated jelly? Should we not expect to find it branching out into species, genera, families, until, by an almost imperceptible bridge, we are carried over into higher Silurian life? But all this, demanded by Darwin's hypothesis, fails in nature: the latest Eozoön differs not from the earliest; it abides in one species, and dies out in the earlier Laurentian. For about

30,000 feet above the latest Eozoön there is no trace of life, the upper Laurentian and Huronian being azoic. Passing on upwards over the 45,000 feet of the Cambrian and its obscure life, which can in no sense be regarded as a development from Eozoön, what do we find in Silurian life? Instead of one species, as Eozoön, and that a mere bit of jelly, we find in the lower Silurian more than three hundred and seventy species. The trilobite is the characteristic type. One hundred and sixty-eight species of trilobites, more than forty genera, spring at once into existence, "having no trilobitic forerunners, and no nearly related preceding organism." And this unheralded trilobite is of great perfection, with an eye of wonderful complexity, and as thoroughly adapted to its method of life as the eagle's eye to-day. All this strongly militates against Darwinism. And among these trilobites, instead of development, degradation in the later Silurian is found: some had lost their eyes, and were otherwise degraded.

Barrande, of Europe, Principal Dawson calls "the first palæozoic palæontologist of our age."

He has made the study of the fossils of the ancient systems of rocks a specialty. He declares that the *facts* he there finds decidedly contradict Darwin's *theory* of evolution by natural selection; and he stands to-day an uncompromising opponent of that hypothesis. Says Dawson, "In connection with his great and classical work on the Silurian fossils of Bohemia, it has been necessary for him to study the similar remains of every other country; and he has used this immense mass of material in preparing statistics of the population of the palæozoic world more perfect than any other naturalist has been able to produce. In previous publications he has applied these statistical results to the elucidation of the history of the oldest group of crustaceans, the trilobites, and the highest group of mollusks, the cephalopods. In his latest memoir of this kind he takes up the brachiopods, or lamp-shells, a group of bivalve shell-fishes, very ancient and very abundantly represented in all the older formations of every part of the world, and which thus affords the most ample material for tracing its evolution, with the least possible difficulty in the nature of 'imperfection of the record.' He claims that the facts of his wide induction prove that variation is not a progressive influence, and that specific distinctions are not dependent on it, but, to use Barrande's words, 'on the sovereign action of one and the same creative cause.' " These conclusions, it is to be observed, are not arrived at by that slap-dash method of mere assertion so often followed by Darwinists, but by the most severe and painstaking induction.

After examining the distribution in time of the genera and species of the brachiopods, Barrande proceeds to consider the animal population of fourteen successive formations included in the Silurian of Bohemia, with reference to the following questions : —

1st. How many species are continued from the previous formation unchanged?

2d. How many may be regarded as modifications of previous species?

3d. How many are migrants from other regions, where they have been known to exist previously?

4th. How many are absolutely new species ?

The total number of species of brachiopods in these fourteen formations is 640, giving an average of 45.71 to each; and the results of accurate study of each species in its characters, its varieties, its geographical and geological range, are expressed in the following short statement, which should somewhat astonish those gentlemen who are so fond of asserting that derivation is *demonstrated* by geological facts : —

First, species continued unchanged ............28 per cent.
Second, species migrated from abroad .......... 7  "    "
Third, species continued with modification ...... 0  "    "
Fourth, new species without known ancestors...65  "    "

This of the brachiopods, Barrande shows holds nearly of the cephalopods and the trilobites; and in fact, that the proportion of species in the successive Silurian faunæ which can be attributed to descent with modification (*i. e.*, to natural selection) is absolutely *nil* — nothing. Barrande may well remark that in the face of such facts, the origin of species is not to be explained by what he terms "the poetic leap of the imagination."

Not only do the early fauna of Silurian times negative Darwinism, but also the fauna of the more recent formations. The elephants and their allies, the deinotheres and mastodons, *e. g.*, appear all at once in the Miocene period and in many countries. The edentates, the rodents, the bats, the manatees, are equally mysterious; and so are the cetaceans, those great mammalian monsters of the deep, which leap into existence in grand and highly developed forms in the Eocene, and which — were Darwinism true — surely should have left in marine deposits some trace of the forms through which, naturally selected, they passed. But even Gaudry, an evolutionist, confesses, "We have questioned these strange and gigantic sovereigns of the Tertiary oceans as to their progenitors, but they leave us without a reply."

And all through the rock record we fail to find those intermediate fossil forms of life, connecting the dying-out and the new-rising creatures, which we should expect to find had all life in a concatenated chain of development arisen out of one primordial form. All the most highly organized groups appear at once and unheralded upon the stage. Darwin recognizes here a difficulty. "Why," says Darwin, "is not every geological

formation and every stratum full of such intermediate links? Geology," he continues, "assuredly
does not reveal any such finely graduated organic
chain. And this, perhaps, is the most obvious
objection to my theory."

And if Darwinism is true in nature, why do
we not find the *now living* creature world a complete jumble of confusion, no well-marked lines
separating the different kinds of animals and vege-
tables, all running into each other in form, habits,
instincts? Our present nature utterly negatives
such a condition of things. Scientists claim that
nature has ever abode the same; nature is uniform   Nature, then, has *never* exhibited such a
jumbling into each other as Darwinism claims;
nature, then, has always negatived Darwinism.

Some evolutionists make much of the gradational forms connecting the reptile and the bird.
But Huxley in his New York lectures confesses
failure here. He says, "If these transitional
forms, which are claimed to link the reptile with
the bird, were the result of development of the
reptile to the bird by natural selection, they ought
to stretch over several geological periods. But

all these gradational forms, and the genuine bird itself, are found in one rock period, are geological contemporaries."

Much has also been made of the horse-like footed animals, found by Prof. Marsh in our own far West, as indicating whence our horse; one of these horse progenitors being only about the size of a fox! The fact that no horses were found on this continent when discovered by Europeans, and no fossil horse bones are found in the rocks, utterly negatives the theory of the creatures discovered by Prof. Marsh being the progenitors of our horse.

The objections just indicated to the acceptance of Darwinism arise when we seek simply to link animal with animal. When we seek to link man with the animals, difficulties intensify. The forms among brutes which most resemble the human are those of the apes; and of these, that of the African gorilla. That there are anatomical resemblances between the gorilla and man is doubtless true; but no less is it true, that the *differences* are so marked that no anatomist claims the gorilla to be the connecting link between the man and the brute. Summing up his comparison of man with the gorilla, Huxley says, "I find that those who attempt to teach what nature so clearly shows in this matter are liable to have their opinions misrepresented and their phraseology garbled, until they seem to say that the structural differences between man and the highest apes are small and insignificant. Let me take this opportunity, then," continues Huxley, " of distinctly asserting, on the contrary, that they are great and significant; that every bone of the gorilla bears marks by which it might be distinguished from the corresponding bone of a man; and that in the present creation at any rate [*i. e.*, among existing creatures], no intermediate link bridges over the gap between man and the apes. It would be no less wrong than absurd to deny the existence of this chasm." And Haeckel, an extreme evolutionist, says ("Creation," II. 277), "I must here point out — what, in fact, is self-evident — that not one of all the still living apes, and consequently not one of the so-called man-like apes, can be the progenitor of the human race. This opinion [of linking man lineally with any of the now living apes], in fact, has never been maintained by thoughtful adherents of the theory of descent; but it has been assigned to them by their thoughtless opponents. The ape-like progenitors of the human race," continues Haeckel, "are long since extinct; we may possibly find their fossil bones in the Tertiary ro ks of Southern Asia or Africa."

The answer of Darwinists to the question " Whence man?" is different from that supposed by many. Darwinists do not hold that man has descended lineally from the apes. They say that away very far back in geological time — probably the Tertiary — existed an animal whence sprang two lines of descent; one of these lines terminates in our now apes, the other in man. This makes the difficulty of deriving man from the brute at all very great; for query, " Where is man's long line of ancestors, reaching from the time of first branch. ing off to perfect man?" — " We cannot tell," answer the evolutionists. Haeckel says he thinks it possible that these ever-developing pithecoid ancestors of ours may be lying buried in the Tertiary rocks of Southern Asia or Africa, or may be in " Lemuria," now under the Indian Ocean !

One of the most repulsive characteristics, to a thoughtful man, in Darwinists is their overweening pride and self-conceit. "We men of science," said Huxley in his New York lectures, "get an awkward habit — no, I will not call it that, for it is a valuable habit — of reasoning, so that we believe nothing unless there be evidence for it; and we have a way of looking upon belief which is not based on evidence as not only illogical but immoral." And with this braggadocio yet on their lips, these same men come to you and me, and say, "The present apes are not our grandfathers. We had grandfathers in a different line of apes, — we guess. These apish grandfathers and grandmothers of ours existed in Tertiary times, — we guess. We have never seen anything that looks like the slightest trace of them, but we guess they existed. The fossil bones of these grandfathers and grandmothers of ours lie deep down in the rocks of Southern Asia, — we guess; if not there, maybe in the rocks of Africa; and if not there, maybe in the old soil of 'Lemuria,' very deep down in the rocks under the Indian Ocean, — we guess. Now, if from us 'men of science,' so exacting in all matters of evidence, you do not accept this as science, you are fogies, — are not up with 'advanced thought.'"

Some seek to link man to ape-like ancestors by means of ancient human skeletons, specially by the crania. They claim that these ancient relics are marked with pithecoid characteristics, and thus help to grade men down to the brute.

Probably the most ancient human cranium yet found is the Engis. But of it Huxley says, "Assuredly there is no mark of degradation about any part of its structure. It is, in fact, a fair average human skull, — which might have belonged to a philosopher, or might have contained the thoughtless brains of a savage."

Quatrefages maintains that no human crania yet found bring us in type nearer the brute, and says, "Believers in pithecoid man must be content to seek him elsewhere than in the only fossil human races with which we are acquainted, and to have recourse to the unknown."

Prof. Virchow, of Berlin, one of the most distinguished living biologists, and an evolutionist, in summing up what has been done in this direction, says, "I should not be surprised if the proof were produced that man had ancestors among the other vertebrates. You are aware that I am now specially engaged in the study of anthropology ; but I am bound to declare that every *positive* advance [*i. e.*, facts] which we have made in the province of prehistoric anthropology has actually removed us farther from the proof of such connection. . . . When we study," continues Virchow, "the most ancient fossil man, who must of course have stood comparatively near our primitive ancestors in the series of ascent, we always find a man just as men are now. On the whole, we must acknowledge that there is a *complete absence* of any fossil type of the lower stage in the development of man. Nay, if we gather together the whole sum of the fossil men hitherto known, and put them parallel with those of the present time, we can assuredly pronounce that there are among living men a much greater number of individuals who show a relatively inferior type than there are among the fossils known up to this time." Prof. Virchow thus declares not only that *facts* to-day do not sustain evolution of man from some lower form, but that the voice of facts is against it. And in his statement that the fossil human crania are on an average higher in grade than those of men living to-day, he accords with the Bible, that the human species has become degraded.

Prof. Asa Gray, in his recent Yale lectures, says, " When the naturalist is asked, 'What and whence the origin of man?' he can only answer in the words of Quatrefages and Virchow, ' We do not know at all.' We have traces of his existence up to, and even anterior to the latest marked climatic change in our temperate zone, but he was then perfected man ; and *no vestige of an earlier form is known.* The believer in direct or special creation is entitled to the advantage which this negative evidence gives."

Thus at present the great chasm separating man anatomically from the brute stands all unbridged.

Not only does man present us new beginnings anatomically, but we find in him other new beginnings : in language, law, social relations, capability of intellectual progression, governance by reason instead of by instinct, moral nature, religious nature.   Here even Huxley acknowledges, "The divergence of the human and simian stirps is immeasurable and practically infinite."

Another line of thought : —

Some creatures there are of very complex adaptations to their peculiar conditions of existence and modes of life ; if any one of these adaptations of instinct, organ, function, is missing, the individual and species must have perished.   These adaptations must have all appeared coetaneously with the individuals of the species.   In the honeybee, for example, " There is an instinct for getting honey ; and answering to this instinct, an instrument just suited for drawing up the honey from the nectaries of the flowers.   There is also a sack for holding it and for producing certain changes in it.   There is an instinct for storing this honey, and a substance secreted, by a peculiar function of the body, that can be moulded into cells to hold it. There are instruments given for using the substance to the best possible advantage, and instincts to guide in the best use, both of the instruments and the substance.   Instinct comes in at the proper place to link all these agencies together.

Let a single link be wanting, and all other parts of the chain are useless as a means of preserving the species; and complicated as this whole process is, it is only a part of the connected series of functional and instinctive adjustments absolutely necessary to honey-bee life, as the species now exists."

A difficulty seemingly insurmountable, here, to the hypothesis that the bee — organs, functions, instincts — arose by the slow process of natural selection !

"Some years ago," says Prof. Schlieden, "one morning I entered, in the lunatic asylum, the room of a madman. I found him crouching down by the stove, watching with close attention a saucepan, the contents of which he was carefully stirring. At the noise of my entrance he turned round, and with a face of the greatest importance, whispered, 'Hush, hush! don't disturb my little pigs; they will be ready directly.' Full of curiosity to know whither his diseased imagination had now led him, I approached nearer. 'You see,' said he, with the curious expression of an alchemist, 'here I have black puddings, pigs' bones, and bristles in the saucepan, — everything that is necessary; we only lack the vital warmth, and the young pig will be ready made again.'"

In connection with this saucepan man, with his "black puddings, pigs' bones, and bristles," looking for his piggy to appear, declaring we have everything that is necessary, — we only lack the "vital warmth," — we may call to mind the Darwinists, who sing in chorus to us, "Hush, hush! don't disturb my little protoplasm, my little monad, my little ascidian, my little monkey; man will be ready directly! You see we have here in the saucepan the primeval fire-mist, the promise and potency of life — organic, intellectual, moral — in a cooled lava ball, spontaneous generation, protoplasm, monad, ascidian, monkey; we season these with large hypotheses, and many large chasms, and frequent large jumping, large credulity, and bold unproven theses, — we only lack quite a variety of indispensables!" Darwinism is forever like the lunatic, looking for his pig to appear; is all right if it only had something it has not got, — something wanting here, something wanting there, something wanting almost everywhere.

Says Tyndall ("Fragments," 154), "In more senses than one, Mr. Darwin has drawn heavily upon the scientific tolerance of his age." So the French Academy of Science, declining to elect Darwin as one of its members, gave as its reason that Darwin is not scientific; that "he has too far sacrificed reason to imagination to deserve a place in the front rank of scientists."

"Science," says Huxley, "must consist of precise knowledge." We have just seen that Darwinism utterly fails to come up to this requisition. "The proper scientific mood is the indicative mood. Science tells what has been, what is, and what shall be. But Mr. Darwin's argument is a continual conjugation of the potential mood. It rings the changes on 'can have been,' 'might have been,' 'would have been,' 'should have been,' until it leaps with a bound into 'must have been.' We are reminded, in fact, by Darwin's method of deriving man from the ape by natural selection, of the famous story which Corporal Trim endeavored in vain to recite to Uncle Toby. 'There was a certain king of Bohemia,' said Trim, 'but in whose reign except his own, I am not able to inform

your Honor.' Uncle Toby was more accommo-
dating than we are able to be from a scientific
point of view. But we recommend the gracious
permission accorded the corporal as a most appro-
priate motto for Darwinian speculations : '"Leave
out the date entirely," said my Uncle Toby.' In
almost similar language, 'There was a certain
monkey,' says Mr. Darwin, — of that he is quite
sure, and he frequently reiterates his assurance,—
'there was a certain monkey; but of what period,
or in what country, or of what shape, except his
own, I am not able to inform my reader.'"

Says Tyndall (probably in one of his "hours of
clearness and vigor"), "What are the core and
essence of this development hypothesis? Strip it
naked, you stand face to face with the theory that
not alone the more ignoble forms of animalculæ
or animal life, not alone the nobler forms of horse
and lion, not alone the wonderful mechanism of
the human body, but that the human mind itself,
intellect, will, and all their phenomena, were once
latent in a fiery cloud [the primitive nebula].
Surely the mere statement of such a notion is more
than a refutation. But the hypothesis as held by

many would probably even assert that at the
present moment all our philosophy, all our
poetry, all our science, and all our art, — Plato,
Raphael, Shakespeare, Newton, — are potential
in the fires of the sun; and that even the unsatis-
fied yearnings in us to know our origin must
have come to us across the ages which separate
the unconscious primeval fiery mist from the
consciousness of to-day. Surely," adds Tyndall,
"these notions represent an absurdity too monstrous to be entertained by any sane mind."
This "development hypothesis," of which Tyndall thus speaks, is Darwinism with its Huxleyan
prolongation backward, — the "expectation" that
all terrestrial organisms have arisen from "not
living matter."

Gray, in his recent Yale lectures, although an
evolutionist, declares that the Darwinan natural-
selection hypothesis fails to explain what we find
in nature. He says, " While I see how variations
of a given organ or structure can be led on to
greater modification, I cannot conceive how non-
existing organs come thus to be, how wholly
new parts are initiated, how anything can be *led*

*on* which is *not there* to be taken hold of. . . . All appears to have come to pass in the course of nature, and therefore under second causes; but what these causes are, or how connected and interfused with the first cause, we know not now, perhaps shall never know. And I cannot help thinking," continues Gray, "that Darwin would agree with me, that the principle of natural selection does not account for it."

And Alfred Russell Wallace, co-propounder with Darwin of the natural-selection hypothesis, after enumerating a large number of facts in nature which this hypothesis cannot account for, declares it insufficient; and says specially of man, "The inference that I would draw from this class of phenomena is, that a superior intelligence has guided the development of man in a definite direction, and for a special purpose, by means of more subtile agencies than we are acquainted with. My theory," continues Wallace, "requires the intervention of some distinct individual intelligence, to aid in the production of what we can hardly avoid considering as the ultimate aim and outcome of all organized existence, — intellectual, ever-advancing, spiritual man."

Darwinism requires so many unproven hypotheses for its support, and so completely fails to explain what we find in nature, that it is not worthy even of provisional acceptance as the key for the solution of the mystery of the *modus operandi* of the rise of terrestrial organisms. Says President Porter, of Yale College, "Juniors frequently are evolutionists, but Seniors usually get over it. I predict that in ten years the theory will be fully exploded." Said Agassiz, just before his death, "I think that careful observers, in view of present facts, will have to acknowledge that our science is not yet ripe for a fair discussion of the origin of organized beings. I hold that this world of ours is not the result of unconscious organic forces, but is the work of an intelligent, conscious power." So speaks science to-day, and so spoke Moses long centuries ago: "In the beginning, an intelligent, conscious power — God — made the heavens and the earth and the sea, and all that therein are." As to the *methods* of this intelligent power's working, men are still groping in darkness.

Egyptian, Epicurean, Lamarckian, Darwinian

hypotheses of evolution thrown out, what shall be
the next phase of this Protean theory, which shall
spring up, say its say, flourish its little day, and
die, with the epitaph, " Our Ancestors' Folly"?

Two mental drifts thrust modern thinking
towards Darwin's conception of the rise of terrestrial life and species.   One of these drifts is indicated by Herbert Spencer; viz., a striving to picture clearly to the mind's eye the *modus operandi*
of the rise of life and species.   "The special creation of p'ants and animals," says Spencer, " seems
a satisfactory hypothesis, until you try and picture
to yourself definitely the process by which one of
them is brought into existence." This difficulty,
some seem to think, is lessened by lessening the
size of the plant or animal originated.   This is
utterly fallacious.   Seek, *e. g.*, to picture Huxley's
hypothetical animated "protoplasm" originating
from " not living matter," that " not living matter "
being operated upon by " physical and chemical
conditions " we in this age of the world know
nothing about; and I doubt whether a person
can "picture " to himself the process of the rise
of life on the earth, any more clearly than by the

special-creation hypothesis; we are thrown into a denser fog. Or seek to "picture" to the mind's eye clearly the coming together of the atoms of Darwin's one little primordial monad, and the " breathing of life into it by the Creator": we are just as much at a loss here for a well-defined mental "picture" as if we should seek ideally to represent the *modus operandi* of the creation of a full-sized horse. Not in the bulk of the life appearing, but in the simple appearance of life at all on the earth, coming out from its lava womb, is where the knot of the problem lies. Lessening the bulk of the embodied life appearing does not one iota lessen the difficulty of forming a clear mental picture of its method of rise. Darwinism has no claim to favor here.

The second mental drift thrusting the to-day thinker towards Darwinism is that so strong at present, and intensified by the recent doctrine of correlation of forces, — viz., the spirit of generalization; the desire to refer all phenomena to uniformly operating, irrational, immanent force in matter, to unify under law. But whether we should accept this dogma, — all force working in the universe of matter is irrational, immanent force, — and carry this back even to that exceptional manifestation of power, the rise of terrestrial life and species, as a prejudging element of our scientific formulating, may justly be questioned. If there is more in the universe than mere matter and immanent, irrational force, genuine science will not without reason eliminate or ignore that "more," in its investigation of the origin of phenomena. Both the original propounders of Darwinism, Darwin and Wallace, concede the existence of this "more"; and Huxley, in his "Lay Sermons," says, "When the materialists stray beyond the borders of their path, and begin to talk about there being nothing else in the universe but matter and force, and necessary laws, and all the rest of their 'grenadiers,' I decline to follow them." Says Spencer, "The hypotheses of special creation and development alike recognize an inscrutable cause of phenomena." As genuine scientists, then, we are not to blink this "more" in formulating of the phenomena of the universe; this "more" may have significance for our formulating. We are to rid our mind of prejudice against its activity in originating phenomena; we are not to prejudge, — it may have been active here, a producing force here. Nor must we prejudge the methods nor extent of its activity; an intelligent force, perhaps, this " more;" and it may have operated through some general law in the introduction and evolution of the creatures of the earth, other than by that of "natural selection," — by some law not yet discovered by us, its discovery waiting a Newton in biology.

It is a mistake to think that if the phenomena of the universe are referred to a free intelligence as cause, this removes uniformity of action as now seen in nature, and makes science an impossibility. This intelligent causal power, ever in itself abiding the same, would, in its ordinary operations, be as thoroughly uniform as mere immanent, irrational force. On special occasions, indeed, this free-working intelligence may exhibit more of its contents than previously manifested, and this wholly unannounced; and just this we find in the rise of life and species in fact: vegetable life, a new, unique thing, rises on our earth; animal life, a new, unique thing, rises on our earth; species rise suddenly; and human life, a new, unique thing, rises suddenly on our earth. I see not how we can explain the facts of terrestrial life as exhibited in the rock record so well as by conceiving of Spencer's "inscrutable" power (by all conceded as working on our globe) as an intelligent, free power. This conception of the all-working force unlocks most satisfactorily the mysteries in the history of terrestrial life. Prof. Le Conte, of California (a scientist), accepts, as the basis of the explanation of the past rise of terrestrial life and species, this intelligent, free power. He claims that if we admit the existence of a Deity at all, the only possible philosophical conception is this: God ever abides in the universe of matter, the sustaining and ultimate force of its continuance and phenomena. This indwelling, all-working, intelligent, free power acts ever uniformly, except in special cases when it acts in higher manifestations. First appears mere matter, then a higher manifestation of the within working power appears in crystallization, a yet higher in vegetable life, higher yet in animal life; and onward now along the line of its operations, rise new species; in man is seen its climax of terrestrial working.

Prof. Le Conte illustrates his conception of this intelligent power's working by a line.* A straight line, continuous for some time, represents the duration of the existence of mere primal undifferentiated matter; a little protuberance — elevation — the beginnings of chemical affinity; then follows a higher for crystallization, — " the first gropings of the so-called ' vital principle,'" as Tyndall designates it; later, and a higher appears for vegetable life; a still higher for animal life; farther on a circle rises above the line, distinct from it, a separate thing lifted out of the mere cause-and-effect region of materialism and instinct. This is personal man, — a new, unique product of the all-producing, intelligent Worker, the highest manifestation of himself, his own likeness and image in domination over the earth and its creatures, in conscious, free, personal individuality, moral nature, knowledge, — the goal towards which, from the first rise above dead matter, each successive upward movement tended all along the geologic æons. The goal reached, there is nothing beyond; the creating power rested, rests to-day, the creative days are closed, — it is Sabbath now.

# CHAPTER VIII.

## ANTIQUITY OF MAN (EVIDENCES OF).

WHAT says the Bible, what says science, as to the time of man's appearance on the earth?

Only the first chapter of Genesis gives any hint on the matter; viz., this, — man appeared on the last creative day. Science is at one with this; says man geologically is very recent, — appears among the latest, perhaps is the latest, of earth's creatures.

The Bible and science so accordant, whence so much dust about the era of man's appearance on the earth?

From two sources : —

*First,* The unfounded assumption of scientists.

*Second,* The unfounded assumptions of Biblicists.

I shall first examine some of the statements of scientists.

Until within thirty years, geologists generally maintained that traces of man were found only in the very latest deposits; *e. g.*, superficial soil, deltas, rock now forming. The finding of articles of his manufacture in cavern deposits, mingling with the bones of the (so-called) extincts, — *e. g.*, the mammoth, cave-bear, cave-lion, — incited suspicion that man had lived longer on the earth than had generally been supposed. The period of these extincts was regarded as much earlier than that usually allotted man; the mingling of his remains and theirs in the same deposits was thought indicative that he had lived in their period, and thus man's period was thrown back.

A second class of facts indicating (as claimed) a high antiquity for man was the finding of isolated human bones in peculiar positions: for instance, the Guadaloupe rock-embedded skeletons; the pelvic bone found at Natchez, on the Mississippi, at the foot of a cliff, loosely mingling with bones of extincts; the bone mentioned by Agassiz as having been found in a Florida coral reef; the skeleton found buried deeply under successive forests at New Orleans; the Abbeville jaw-bone; the skull found in California deeply embed-ded in gold drift, under basalt rock.

Again, "Stretching along the sides of many river valleys, both in this country and in Europe, are certain deposits of sand, clay, and gravel, sometimes more than a hundred feet, seldom less than forty feet above the level of existing streams. These terraces have been long known to contain relics of man. In the gravels of the Ouse and Waveney, England, of the Seine, France, but specially of the Somme, Picardy, have been found flint chips, arrow heads, hatchets, in the same layers with bones of the mammoth and other extincts." All these river terraces belong to one period, a period claimed to be much prior to that formerly allotted man.

These three classes of data — viz , cavern de-posits, isolated human bones, river terraces — are the main grounds upon which geologists base their claim for man's high antiquity.

Let us interrogate these three witnesses.

*First*, Isolated bones.

The isolated-bone proof of man's great antiquity is by geologists regarded the least reliable. It is now on all hands conceded that unwarranted antiquity has been claimed for some of these bones. For instance, the fossil Gaudaloupe man, referred to a great antiquity by Nott and Gliddon, is now, by Prof. Dana, declared to be a Carib Indian, and not more than two or three centuries old. So completely is the claim of great antiquity for these skeletons abandoned now, that they are never in our day even mentioned in treatises on the antiquity of man.

Agassiz, by estimates from the present rates of deposition of coral, claimed for the Florida bone an antiquity of 14,000 years. But Count Portales, its finder, now declares the bone was not found in coral rock at all, " but in a fresh-water sandstone, on the shores of Lake Monroe, Florida, associated with fresh-water shells of species still living in the lake; and no date can be assigned for the formation of that deposit."

The Natchez bone, Lyell, in his second visit to America, pronounced of little value. Says Winchell, "From being the relic of a preglacial man, it suddenly became the bone of a red Indian, perhaps one hundred and fifty years old."

The Abbeville jaw-bone is a bone of contention to scientists themselves, and we may therefore count it out.

The New Orleans skeleton, buried under sixteen feet of river mud and four successive cypress forests, Dr. Dowler estimated was 57,-000 years old; Lyell approves the estimate. On the other hand, the United States army engineers, Humphreys and Abbot, claim that the ground on which New Orleans stands, down to the depth of forty feet, instead of sixteen feet, has been deposited within 4,400 years; and Dr. Foster says that the superimposed wood, regarded by Dr. Dowler as four successive cypress forests, "may be nothing more than drift-wood brought down by the river, which became embedded in the sediments." Says the geologist Fountaine, "The Mississippi sixty years ago used to flow where now Tchoupitoulas Street is, in the heart of the business part of the city, a quarter of a mile from the present shore. The river at New Orleans is from one hundred and sixty-two to one hundred and eighty-seven feet deep. By under-mining and engulfing its banks, with everything upon them, logs tangled in vines and bedded in mud, cypress stumps, Indian graves, and modern works of art are suddenly swallowed up and buried, at all depths from ten to one hundred and eighty-seven feet." In view of all this, a recent writer on the New Orleans skeleton says, "To claim 50,000 years for it is provocative of laughter."

Some Californian workmen a few years ago presented Prof. Whitney, State geologist, with a skull claimed to have been found one hundred and thirty feet below the surface, covered by basalt rock. The skull was heralded in this country and in Europe as a veritable preadamite skull, lying geologically periods lower than any other yet discovered human relic. It appears now that the workmen hoaxed the professor as to where they found the skull. Even Quatrefages says, "The most serious doubts exist as to the antiquity of the specimen, which seems to have been found in disturbed ground."

The isolated-bone proof of man's great antiquity thus utterly breaks down. Geologists speak now little of it.

*Second*, Cavern deposits.

We can tell nothing from the caverns themselves, nor from their deposits, as to their geologic period. The special argument urged for the great antiquity of their human relics is the mingling of these relics with the bones of the extincts; and here, I claim, is one of the unwarranted assumptions of scientists. Instead of making this mingling and coexistence of man and the mammoth a reason for immediately thrusting man back a geological period earlier than previously allotted him, into the mammoth period, why is not this just as good cause, this mingling of relics, for bringing the mammoth and his congeners forward a geological period into the period of man, and thus make the epoch of the mammoth greatly less distant from our day than we had previously supposed it?

The key to the problem of antiquity of the human relics of the caverns lies in the answer to the query, When did the mammoth and his fellows live?

*Third*, And as the terrace problem now stands, its only proof of man's great antiquity, is this same mingling in these terrace deposits of human relics with the bones of the extincts. For the solution, therefore, of both the cavern and the terrace problems, I shall now address myself to the solution of the query, When did the mammoth and his fellows live?

The old tenet of the great antiquity of the mammoth epoch is being examined in these later years, and a great reaction is taking place; geologists now bring down the mammoth to a comparatively recent date. Messrs. Prestwich and Falconer, of England, began this work twelve years ago. They brought forward the cavern deposits containing bones of the extincts an entire geological period, from the Bowlder Clay to the post-Pliocene; and the justness of this is now by geologists conceded.

The remains of the mammoth in America are found in superficial deposits, in situation and condition of preservation indicative of their comparatively recent extinction. "Almost any swampy bit of ground," says Prof. Shaler, "in Ohio or Kentucky contains traces of the mammoth. At Big Bone Lick, Kentucky, the remains are so well preserved, as to seem not much more ancient than the buffalo bones which are found above them." Prof. Winchell says he has "seen the bones of the mammoth embedded in peat, at depths so shallow that he could readily believe the animals to have occupied the country during its possession by the Indians." Says Lyell, "That the mammoth was exterminated by the arrows of the Indian hunters is the first idea presented to the mind of almost every naturalist." In accordance with this view of Winchell and Lyell, we find an Indian tradition of the existence of the mastodon, seemingly earlier extinct than the mammoth: "that they were often seen, that they fed on the boughs of a species of lime-tree, and that to sleep, they did not lie down, but leaned against a tree."

In Europe, also, bones of the mammoth have been found in peat deposits, — one of the most recent of all superficial accumulations, and even now forming. For instance, two perfect heads of the mammoth were brought to light by excavations made for a railway in 1847, at Holyhead, England. They were found in a bed of peat three foet thick. Lyell thinks that these individuals must have perished at a comparatively recent date.

And the tendency at present is to bring down much nearer our own time than formerly all the (so-called) extincts of the cavern and terrace epochs; some are even identified with species now living. Having thoroughly canvassed the evidence, and summing up on this matter, Southall says, "The cave-horse, the cave-bear, the cave-lion, the cave-hyæna, are still living; the cave-lion is mentioned historically even in Europe a few centuries before our era; wild horses scoured the plains of Russia a few centuries ago; the urus survived to the sixteenth century, the aurochs still survives in Russia; the reindeer is traced down in Europe to the twelfth century; the great elk survived equally as late; the mastodon, mammoth, and woolly rhinoceros are found under circumstances that imply their existence only a few thousand years ago."

Such is the present condition of the problem of the epoch of the (so-called) extincts of the caverns and terraces, with whose bones human relics mingle, and which mingling is the grand stock argument, in these latter years, for man's great antiquity. In the light of recent discoveries and readjustments of data, this claimed great antiquity dwindles down to a few thousand years. No need to thrust man back tens of thousands of years, from any data given us in cavern or terrace deposits, — only a very few thousand years answers every demand.

At the meeting of the British Association for the Promotion of Science, last year, 1879, Dawkins, a very high authority in these matters, said, "There is no proof that the animals with whose remains man's relics are commingled are of extreme antiquity." Prof. Dawson ("Earth and Man," 295) claims that the St. Acheul gravels of the Somme can "scarcely exceed 4,000 years."

## CHAPTER IX.

THE ANTIQUITY OF MAN (CONTINUED).    YEAR
MEASURE IN GEOLOGY.

In the attempt to measure in years the interval
between us and the terrace epoch, by geological
data, appear some of the ungrounded assumptions
of scientists.

Says Lyell, "The terraces are post-Pliocene;
intervening between us and that is the Recent, —
the epoch of the deltas, peat, Swiss lake villages.   Determine how many years have been
occupied in the formation of the deltas or peat, or
the era of the Swiss lake villages, — then, yet
back of that lived the terrace man."

Further, "Some post-Tertiary deposits, on the
coast of Norway, of marine formation, are now "
(Lyell claims) "elevated above the sea six hundred feet.   To attain such elevation, allowing

two and a half feet rise per century" (which Lyell regards large), "requires 24,000 years." He claims this rise has mostly taken place since the terrace period.

I shall examine Lyell's data.

*First,* Deltas.

The Mississippi delta is the one of which most use has been made in estimating time. Lyell demands for the formation of this delta "probably more than 100,000 years." On the contrary, Messrs. Humphreys and Abbot, of the government survey, have thoroughly explored it, and demand for its formation only 4,400 years; and it is doubtful if even this diminished time is required. Singular elevations at the bottom of the gulf near the delta, perhaps in the delta itself, known as "mud lumps," occur with frequency, by means of which acres in extent are sometimes raised, often above the surface of the water. Allowance must be made, shortening indefinitely the estimate of 4,400 years, for the material added to the delta by these "mud lumps."

The eminent French naturalists, Dolomieu, Cuvier, and Beaumont, claim that a few thousand years suffice to form all the deltas in the world. Beaumont estimates the Mississippi delta at 1,300 years.

These discrepancies of estimates among scientists declare very emphatically that they are themselves utterly afloat, and their estimates in years of the antiquity of man from delta data wholly unreliable.

*Second*, Swiss lake villages.

Geologists have sought to determine approximately, in years, from estimates on present lake depositions, the era of the submerged villages. These estimates are as little reliable as those of the deltas. But the extreme antiquity claimed for these villages is not above 3,000 or 4,000 years; we may therefore pass them.

*Third*, Peat formations.

The French naturalist, Perthes, estimates the rate of the growth of peat anciently in the Somme Valley at one inch a century. The deposit is thirty feet deep, and at Perthes's estimate demands 30,000 years. But before this peat began to be laid down, the flint-knife men of the terraces were fishing in the river. This long outlook into the past, forced upon us by Perthes's estimate, even Lyell hesitates to accept. Facts utterly refute it. In the peat are trunks of trees standing erect where they grew, birches and alders three feet high, also prostrate trunks of oaks four feet in diameter. According to Perthes's estimate, it required a century to cover up these stumps and prostrate trunks one inch; but at the end of the century, where would have been all the uncovered thirty-five inches of the stumps, the forty-seven inches of the trunks? Rotted off. The estimate is absurd. It would have required 3,600 years to cover the stumps, 4,800 years to cover the trunks: neither stumps nor trunks would have waited so long. On the contrary, in Irish bogs an increase of peat of two inches a year has been observed, — sixteen feet instead of an inch a century. Two centuries would thus lay down all the peat in the Somme Valley. And the Frenchman D'Archiac tells us that it is estimated that in the upper valley of the Somme, peat to-day grows at the rate of eleven and one half feet a century; this rate of deposition would require only three centuries for the work for which **Perthes** demanded 36,000 years. But the fact is, the peat growth depends só much on attending conditions — the kind of peat, its place of growth, the climate — that peat depositions give us no reliable data for year-measure estimates; it is absolutely of no value as datum in estimating in years man's antiquity. A boat containing bricks was found under the lowest layer of the Somme peat, but there were no bricks in Gaul till the Roman era.

*Fourth*, Elevated Norway formations.

Lyell makes an attempt to measure in years the recent period by estimates on the coast of Norway; he claims a present rate of rise of two and one half feet a century. He finds shells of the post-Tertiary period six hundred feet above the level of the sea, requiring to raise them to such elevation, at the present rate of rise, 24,000 years. But at the beginning of this period the terrace man lived.

Since Lyell's statement, Prof. Kjerulf, of Christiana, has made the government geological survey of the Norway coast, and carefully examined the raised beaches and terraces, and declares

Lyell's statement to be utterly without foundation.

*First*, he says the uppermost limit of post-Tertiary rise is only sixty instead of six hundred feet, reducing Lyell's 24,000 to 2,400 years.

*Second*, the abrupt edges of the terraces, separated by level areas, indicate sudden elevations, succeeded by periods of rest, utterly destroying all data for computation of time.

*Third*, he declares that the coast is not now rising, but that this is a stationary period; Lyell's two and one half feet rise per century, the basis of his entire calculation, is a myth.

Thus bursts, at the touch of the finger of a more exact science, Lyell's 24,000-years Norway bubble!

By this hasty glance at the most trusted data by which men of science have attempted to estimate *in years* the era of man's introduction upon the earth, we see that all is simply guess: there is nothing assured, reliable; it is not science, — precise knowledge. And we behold again on the plains of guess what was

anciently seen on the plains of Shinar, — a
Babel of confusion, men using a gibberish which
to each other conveys no knowledge, and from
which nobody else can gather anything but confusion.

# CHAPTER X.

### THE ANTIQUITY OF MAN (CONTINUED). BIBLICAL CHRONOLOGY.

THEOLOGIANS have customarily claimed that the Bible gives us data by which we may determine with close approximation, in years, the era of man's appearance on the earth; also, that Usher's date is the Bible date. I deny both.

That the Bible gives us no data for estimating man's era with certitude of a close approximation, and that Usher's date is a mere human estimate, wholly unreliable, a statement of the " Oxford Chronological Tables " indicates. Say these tables, " Chronologers have piled system upon system, without adding much to our stock of knowledge respecting the remote ages of antiquity. Thus, for example, there are not less than three hundred different dates assigned as the era of creation, varying, in their extremes, not less than 3,000 years."

I have examined a list of one hundred and twenty different estimates from Bible data of man's era, by as many different scholars, each estimate different from all the others; the extremes, 6,984 B. C. and 3,483 B. C., — one estimate more than double that of the other; and intervening between these were the one hundred and eighteen other estimates. This clearly declares, as also the "Oxford Tables," that the Bible gives us no data for estimating with close approximation, in years, man's era.

Query, then, what becomes of the Biblical genealogical tables carrying us back to Adam?

There comes of them all that ever was intended to come of them, all that ever legitimately can come of them, viz., ability to trace family descent. Thus far they are reliable; but use them as exact data for chronological estimates, they are used for a purpose for which they were never intended — are wholly unfit.

Says Pritchard, "The omission of some generations in Oriental genealogies is a very ordi-

nary thing, the object of the genealogy being
sufficiently answered by inserting only the *con-
spicuous and celebrated* names which connect the
individual with his remote ancestry." Eichhorn
and Michaelis note the same. This sets us utterly
afloat! Who will tell us where the omissions are
in the long genealogical lists of Genesis, and how
many centuries these omissions represent?

Further, " The Samaritan Bible has a different
set of dates from the Hebrew copies, and both
from the Septuagint, and all these from the Ethi-
opic version; and this not merely in one text, but
the discrepancy runs through nearly the entire
genealogy. The Hebrew, Samaritan, and Septuagint versions, in giving the ages of the patri-
archs before Abraham, vary in the aggregate
about 1,500 years."

On the whole matter of Bible chronology,
Pritchard says, " The Hebrew chronology may be
computed with accuracy to the era of the building
of the Temple, or at least to the division of the
tribes, — tenth century B. C. In the interval
between that date and the arrival of Abraham in
Palestine, Hebrew chronology cannot be ascer- .

tained with exactness, but may be computed with near approximation to the truth. Beyond Abraham, we can never know how many centuries, nor even how many chiliads of years, may have elapsed since the first man of clay received the image of God and the breath of life. Still, as the thread of genealogy has been traced, though probably with many and great intervals, the whole duration of time from the beginning must apparently have been within moderate bounds, and by no means so wide and vast as the Indian and Egyptian fabulists assert." Pritchard might now have added, "some geological fabulists assert."

Says Bunsen, "The study of the Scriptures has long convinced me that there is no connected chronology prior to Solomon."

Says Conant, "I do not think we have exact and full data for determining with absolute certainty the number of years from Adam to Abraham."

I regard these statements of Pritchard, Bunsen, and Conant the correct view of early Bible chronology; viz., the Bible does not give us data from which with certainty we can determine the length of the period intervening between Adam and Abraham. Pritchard's other statement I regard also correct; viz., the Bible genealogies impress us with the idea that the whole duration of man's existence upon the earth is contained within moderate limits. That this is so, the recentness of the rise of the arts and sciences in their fulness indicates; as also the narrow limits of all assured national chronologies: *e. g.*, India, twelfth century B. C.; Assyria, sixteenth century B. C.; Babylonia or Chaldea, twenty-second century B. C.; China, twenty-fourth century B. C.; Tyre, twenty-seventh century B. C.; Egypt (the utmost claim of Lepsius for its monuments), thirty-fifth century B. C. All these come within Usher's date for Adam, 4,000 B. C. But even these dates, contracted as they are, are by no means proven. Says the Egyptologist, Wilkinson, "No certain era has been established in early Egyptian chronology." Says Lyell ("Antiquity of Man," 380), "True history and chronology are the creation, as it were, of yesterday. Thus the first Olympiad is generally regarded as the earliest date on which we can rely in the past annals of mankind, — only 776 B. C.; and no ancient monuments and inscriptions seem to claim a higher antiquity than fifteen centuries before Christ."

# CHAPTER XI.

THE ANTIQUITY OF MAN (CONCLUDED). PRESENT
CONDITION OF THE PROBLEM.

THERE is at present a decided and growing
tendency among scientists to contract greatly
man's antiquity. We see this tendency in what
Huxley said at the last year's (1879) meeting of
the British Association for the Promotion of
Science. After the president of the Anthropolo-
gical Section, Dr. Tylor, had read a paper claiming a high antiquity for man, urging the stock
argument, the stone implements of the Somme
terraces, Prof. Huxley rose and said, "I have a
warning for anthropologists. Few are aware of
the immense changes that have taken place in the
geological formations of large parts of Europe
during the historic period. No one can show that
similar changes may not have taken place with

as great rapidity in the Somme Valley; and the implements found in the terraces cannot be relied on to prove a great antiquity of man." This coming in such a place, from such a man as Prof. Huxley, is highly significant.

It has already been shown in this volume that evidence of man's antiquity from cavern and terrace deposits dwindles down to a very few thousand years.

The tendency of science in recent years is also greatly to contract the time intervening between us and the glacial epoch. The glacial epoch has been set back from our day as far as 1,280,000,-000 years. Lyell fixed it prior to us 800,000 years; later, he said 200,000 years.

The determination of the time intervening between us and the close of the glacial epoch is a vital element in the solution of the problem of the antiquity of man, as that problem *in attained facts* now stands. All relics of man yet discovered, whether in America or Europe, lie on this side of the glacial or drift epoch; determine when that epoch closed, and we can say, As *facts* now stand, man did not reach back of that date; he might have appeared many years on this side of it.

There is but little evidence of man's great antiquity in America. The stone implements found in California in auriferous gravels covered with basalt rock, however deep, give us no proof of immense antiquity for man, as these basaltic outflows are recent phenomena, — are post-glacial.

Geologists relegate the cavern and terrace deposits to the same geologic period. Lyell declares all these deposits bearing human relics to be post-glacial, whether found in England or elsewhere. These terrace deposits in England are so related to contiguous glacial deposits that indisputably they are the later formation, — are post-glacial; *e. g.*, the Ouse at Bedford, the Hoxne at Diss (Lyell, "Antiquity," 164, 166). And of the Somme Valley gravels, Winchell says ("Pre-Adamites," 425), "It appears to demonstration that the entire river valley was excavated after the glacial drift was laid down. The valley is cut through the glacial drift and into the chalk. But the flint-bearing gravels are still more recent, having been deposited along the chalk slopes of the valley." As Winchell here allows a period to intervene between the drift and the laying down of the relic-bearing gravels of the terraces long enough to erode the drift and chalk, so Prof. Dana claims that a short geological period, which he names the "Champlain," intervened between the glacial and terrace periods ("Geology," 558).

Principal Dawson, as Lyell, makes all the cavern and terrace deposits post-glacial ("Earth and Man," 283). The British archæologist, Evans, at the 1877 meeting of the Geological Society, said he had not met with any evidence of man's presence in glacial or pre-glacial times. Says Winchell ("Pre-Adamites," 425), "The *general tenor* of the evidence connected with the occurrence of human remains proclaims everywhere the advent of man in Europe to have been subsequent to the general glaciation; but it happened during the progress of the disappearance of the glaciers. He was an inhabitant of France early in the Champlain period, while the rivers were still swollen from the melting snows."

When I say that man is "post-glacial," it is in the sense of being posterior to the grand submergence of the land in the waters of the oceans which took place both in America and Europe (breaking up the grand ice sheet so extensively covering the land as an encasement during the glacial period proper), and its subsequent elevation; local glaciers on a contracted scale yet remaining.

Even the extremist Haeckel, in his *conjectures*, claims no antiquity for language-speaking man higher than the glacial epoch ("Creation," II. 18). And Prof. Dawkins, at the meeting of the British Association, 1878, said, "There is no proof of any kind that man is pre-glacial"; with which Huxley at the same meeting concurred, adding, "And recollect that drift is only the scum of the earth's surface."

Geikie, indeed, in his "Great Ice Age," puts forth claims for man's pre-glacial existence, but gives us no *facts* to sustain his position, except that a human fibula had just been found at the writing of his book (1873), under "stiff glacial clay." The "human fibula" has recently, before the British Association, by Prof. Busk, its discoverer, been declared the bone of a bear!

In science the matter of man's antiquity to-day stands thus: Some extremists claim for man an existence during the glacial epoch; authority and facts declare all relics of man post-glacial.

The great question thus is: When closed the glacial epoch, — how long since the Drift?

Prof. Andrews, of Chicago, from observations on our northern lakes, declares his belief that " the total time of all the deposits since the glacial epoch is somewhere between 5,300 and 7,500 years."

Careful observations have been made by Prof. Winchell at St. Anthony's Falls, on the Mississippi, on the length of the chasm worn and the present rate of wear. The mean result of the different estimates for the close of the glacial epoch was 8,859 years. These observations are considered by some the most reliable yet made in attempts to solve this problem.

French savants have been seeking a solution of this problem. M. de Ferry, from observations on the deposits of the Saône, gives an interval

from our day to the glacial epoch of 9,000 or
10,000 years. M. Arcelin, from estimates on the
same data, makes this interval 6,750 years. M.
Kerviler, taking as data the alluvial deposits of the
Loire, makes the interval 6,000 years. Of this
latter estimate Quatrefages says, " This represents
a very moderate antiquity, and corresponds almost
entirely with the dates of Manetho." Upon a review of the entire facts yet attained by science
auxiliary to the solution of this problem, we have
this : All relics of man yet found are this side the
glacial epoch; the interval separating us from
the glacial epoch is from 6,000 to 10,000 years;
the flint-implement man of the terraces may not
have appeared immediately at the close of the
glacial epoch, — Dana claims the " Champlain "
epoch intervened.

In his work just issued ("Pre-Adamites," 473,
431), Winchell says, "Prehistoric Europeans, so
far as inductively known, were post-glacial; I discover no valid ground whatever for the opinion
that the Stone Age in Europe began more than
2,500 or 3,000 years B. C."

Man did not originate in Europe, and we must

allow some time for the migration of the Asiatic man from his primitive habitat, before he began chipping flint in the valley of the Somme. This time must be added to the age of the flint implements, but this does not necessarily demand a very protracted period.

These latest and most reliable utterances of science as to traces of man's appearance on the earth, — how like the utterance of the Bible, so far as we may venture to conjecture anything from its data! I have already indicated the unreliability of estimates in year measure, both in geology and early Bible chronology; but taking the most reliable estimates in both these provinces for what they may be worth, they strikingly harmonize. Says science, "Not earlier than from 6,000 to 10,000 years prior to the present day do I find any trace of man on the earth; from my data he cannot have appeared earlier, — he may have appeared later." The Septuagint (Mai's edition) makes Adam's date from our day 7,411 years; Hebrew Bible, 5,945 years; another Biblical estimate gives us 8,863 years. (See a very much more liberal Biblical chronology, as estimated by Mr. Crawford,

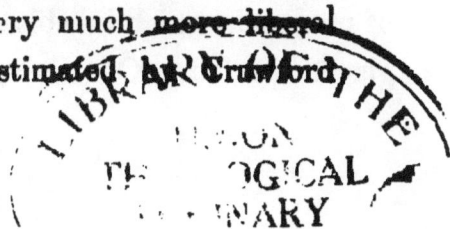

("Patriarchal Dynasties"), quoted by Winchell ("Pre-Adamites," 449) : the period between Abraham and Adam is 10,500 years; Adam's date, 14,381 B. C.)

This very moderate antiquity of man, and the fact that the earliest fossil man does not differ in species nor variety from man of the present, militate strongly against Darwinism. Dawson ("Earth and Man," 356), after saying, "Man is a very recent animal, dating no further back in geological time than the post-glacial," says, "Further, inasmuch as the oldest known remains of man occur along with those of animals which still exist, and the majority of which are probably not of older date, there is but slight probability that any much older human remains will ever be found. The Engis skull is perhaps the oldest known human skull. This fossil Belgian man is believed to have lived before the mammoth and the cave-bear had passed away, yet he does not belong to an extinct species or even variety of man."

Further, Pictet catalogues ninety-eight species of mammals which inhabited Europe in the post-glacial period. Of these, fifty-seven still exist unchanged, and the remainder have disappeared. Not one can be shown to have been modified into a new form, though some of them have been obliged, by changes of temperature and other conditions, to remove into distant and now widely separated regions. Further, it would seem that all the existing European mammals extended back in geological time at least as far as man, so that since the post-glacial period no new species have been introduced in any way. Here we have a series of facts of the most profound significance. Fifty-seven parallel lines of descent have in Europe run on along with man, from the post-glacial period, without change or material modification of any kind. Some of them extend without change even further back. Thus man and his companion mammals present a series of lines, not converging as if they pointed to some common progenitor, but strictly parallel to each other. In other words, if they are derived forms, their point of derivation from a common type is pushed back infinitely in geological time; but negativing such primeval derivation of man from some allied animal, is the fact that so early as in post-glacial times all trace

of man ends, and no trace of any previously existing creature from which even extremists would
derive him is found. But the absolute duration
of the human species does not affect the argument.
If man has existed only six or seven thousand
years, still at the beginning of his existence — so
far as we have found his remains — he was as distinct from lower animals as he is now, and shows
no signs of gradation into other forms. If he has
really endured since the great glacial period, and
is to be regarded as a species of a hundred thousand years' continuance, and we relegate the Engis
skull to that date, still the fact is the same, and
is, if possible, less favorable to derivation. In
such case we ought to find the Engis skull decidedly nearing in type our brute ancestor, which it
decidedly does not. If 100,000 years ago the individual of the human species was nothing less
than a man, we may thoroughly believe that in
his earth-history he was never less than a man, —
man as we now see him.

# CHAPTER XII.

## UNITY OF ORIGIN OF THE HUMAN SPECIES.

The old din about the unity of species of the human races has passed away : this is now conceded. But some are still inclined to deny the origin of mankind from one pair, — deny the *unity of origin* of the species. Says Haeckel ("History of Creation," 304), "This supposition (unity of origin), which our modern Indo-Germanic culture has taken from the Semitic myth of the Mosaic history of creation, is by no means tenable."

Other scientists claim that all human races have risen from a single pair. Says Quatrefages ("Species," 84), "What science may affirm is that *from all appearances* each species has had, as point of departure, a single primitive pair"; and he says further, "No facts have as yet been discovered which authorize us to place the cradle of the human race elsewhere than in Asia." So Prof. Winchell (" Pre-Adamites," 409) maintains that all races of men are derived from "a single original centre of humanity."

The universality of the tradition of the flood indicates a common origin of all races of men from the flood-saved family of Noah. The tradition is found among the Hindoos, Chaldeans, Egyptians, Chinese, Persians, Greeks, Americans. The American Arctic Crees had a tradition of a universal deluge, from which one family alone escaped, with all kinds of beasts and birds, on a huge raft. The Mexican Noah was named Coxcox. He and his wife were all that escaped of the human race. He sent forth, as the waters assuaged, several birds; only one, the hummingbird, returned, holding in its beak a branch covered with leaves. One of the traditions of the Mexicans deduced their descent and that of the tribes from seven persons, who, after the deluge, came forth from as many caves. The Persians had a similar version of the story. The universality of this tradition, and its similarity in particulars among the peoples of mankind, strongly point to a unity of origin of all these peoples.

And it is a canon in science that causes are not to be multiplied unnecessarily. Science to-day declares that one pair is sufficient cause to produce all human races; then why unphilosophically multiply centres?

Comparative philology strongly sustains the hypothesis of the unity of origin of all human races. It is now a theorem in comparative philology that the more critically languages are investigated in their roots, vocabularies, and grammatical forms, the more do apparently isolated languages become referable to a common stem, — the more do all the languages of mankind point back to a unity of origin.

Dr. Latham claims that "the laws of language are the laws of growth and development. It seems that a period wherein no inflections are evolved precedes the period of inflections."

Philologists incline to the opinion that monosyllabic ideographic Chinese is of all existing languages nearest the primitive type of human speech, and that from some form similar to monosyllabic Chinese, by a natural growth, all other languages are derived. Philologists divide the languages of mankind, according to their development in inflections, etc., into three great classes; viz., Turanian, Semitic, Aryan or Indo-European.

The Turanian area, commencing in the north-east of Europe, embraces Lapland and Finland, reaches across the North of Russia, takes the Ural Mountains and the Caspian Sea for its western limit; for its southern limit, the mountain range eastward from the Caspian to Hindoo Koosh, thence along the Himalayas down to the eastern shore of the Bay of Bengal; from these limits it includes all Northern Asia, monosyllabic China and Thibet excepted. This class, in simplicity of structure and in want of real inflections, approaches in some degree the monosyllabic idioms.

The Semitic area embraces Syria, Mesopotamia, part of Babylonia, the Arabian peninsula, ancient Egypt. The African Semitic Galla tongue, south of the equator, dovetails into the Kaffir dialects, which are again affined to the southwestern languages of Africa.

The Indo-European area includes nearly all Europe, and a portion of Southwestern Asia. Modern philology points to ancient Asiatic Bactria as the old seat of the common mother-language of this Indo-European class, of many sisters. Philology thus places the fountain-head of these many linguistic streams in the same region as the Bible does the Babel of the dispersion.

The Oceanican and American languages are affined to the Turanian class; Turanian agglutination marks the American.

The Chinese and Thibetan are, through the monosyllabic Tai of the Turanian, and the non-monosyllabic Bhotiya of the same class, affined to the Turanian.

These three great classes of human languages and their affined tongues cover nearly the entire earth.

If there can be shown to have been a primitive connection between these three all-embracing classes of human speech, this would indicate a probability of a primitive connection, — one common origin of all the peoples whose speech those languages are, — viz , all mankind. And just this primitive connection philology now claims to be able to show.

For instance, Lepsius, Delitzsch, and Max Müller claim a common element in Sanskrit (of the Aryan class) and Semitic. Says Müller, " There is even now sufficient evidence with regard to a radical community between Aryan and Semitic dialects, to enable us to say that their *common origin* is not only possible, but as far as linguistic evidence goes, probable."

Again, in the least developed form of Semitism, Khamitism (the idiom of the cuneiform inscriptions, and of ancient Egypt), we find Aryan affinities, also Turanian, and the form of monosyllabic Chinese; and in the least developed form of the Aryan class, Celtic, we find affinities with the Semitic and Turanian classes. This running into each other of the most ancient and least developed forms of the three great branches of human speech, and their nearing (as in Khamitism) undeveloped monosyllabic Chinese, is just what we might have expected to find had all human language sprung from some common source, and had each great emigration carried away with it

its heirloom of speech from the one homestead,
developing its primitive material according to its
own idiosyncrasy and needs.

Says Klaproth, "The universal affinity of language is placed in so strong a light, that it must
be considered by all as completely demonstrated." Says Herder (a rejecter of the divine
origin of the Mosaic record), "The human race,
and language therewith, go back to one common
stock, to a first man, and not to several dis-
persed in several parts of the world." Says
Schlegel, "Although many are the nations of the
earth, their languages are evidently nearer or
more distant varieties of a single mother tongue,
spoken by one family of people; which proves that
in the distant and indeterminate antiquity, emigration took place over wide tracts of country, from
a common original abode. This is no hypothesis,
but a fact clearly made out."

Says Max Müller, "And now, as we gaze from
our British shores over that vast ocean of human
speech, with its waves rolling on from continent
to continent, rising under the fresh breezes of the
morning of history, if we hearken to the strange

sounds rushing past our ears in unbroken strains, it seems no longer a wild tumult, an ἀνήριθμον γέλασμα, but we feel as if placed within some ancient cathedral, listening to a chorus of innumerable voices; and the more intensely we listen, the more all discords melt away into higher harmonies, till at last we hear but one majestic trichord, or a mighty unison as at the end of a sacred symphony. Such visions will sometimes float through the study of the grammarian, and in the midst of toilsome researches his heart will suddenly beat, as he feels the conviction growing upon him that men are brethren in the simplest sense of the word, — the children of the same father, — whatever their country, their color, their language, or their faith."

# CHAPTER XIII.

## FINAL DESTINY OF THE EARTH.

(*a.*) THE idea of cycles has been a favorite among men when conceiving of the history of our globe. This was the Hindoo idea; this also gleams through Egyptian and Greek transmigration. Says science, " Our globe was once a burning lava ball." Says the Bible, " Our globe is a circle-walker, — is moving round the circle to the point and condition from which it set out, a burning lava ball; the elements of the earth shall melt; the ascending smoke, vapors, gases, shall hide the heavens, cause them to disappear as a scroll rolled up." Such is Biblical prophecy : what says science of its probability or possibility?

Science says, " No atom of matter is annihilated." Our globe, then, and its surrounding atmosphere contain precisely the same elements

as when the earth was a lava ball; it only needs
these elements to come into their old relations
that all be a lava ball again. And it may be that
all the elements of nature in the midst of which
we live are in one grand march, fixedly moving
around the circle to this very point, and "in a
moment, in the twinkling of an eye," all to be
a fiery furnace again.

Science knows full well the inflammable nature
of the elements constituting the earth and its
atmosphere. Sir Charles Lyell quotes approvingly the thought of Pliny, "It is an amazement
that our world, so full of combustible elements,
stands a moment unexploded."

What is that which sustains the flame of the
candle, the rolling fire-waves of the burning building, the intense heat of the smelting furnace trans-
muting iron into a red, glowing liquid stream?
A gas called oxygen, the base of all combustion; combustion is the product of oxygen in
combination with any other substance. When
the process of combustion is conducted in pure
oxygen, even some of the metals commonly
regarded as incombustible may be made to burn

with wonderful brilliancy. A steel wire, for instance, gives out a bright flame.

How awfully intense must be the heat of a furnace whose fuel is pure oxygen! Just such a possible furnace is our world, over which is written, "Reserved for fire."

Our atmosphere is about forty-five miles high, composed essentially of two gases, nitrogen and oxygen. Oxygen in volume is one fifth of this forty-five miles wrappage. Let the atmosphere by some shock be suddenly decomposed, and oxygen, the heavier gas, fall to the earth; a wrappage nine miles in thickness of this fearful furnace gas would encircle this globe. Who can set limits to the melting powers of such a mass of this furnace fuel?

Again, water is constituted of oxygen and hydrogen; by weight, oxygen 8 parts, hydrogen 1. The atmosphere is impregnated with aqueous vapor to the height of five or eight miles. There is water enough in the oceans, lakes, rivers, to cover the entire globe two miles ("Challenger") in depth. Immense quantities of water are within the earth: subterranean rivers and lakes, and that

mass in stationary collections in gravelly and other
loose strata, from any of which, on perforation,
come the ordinary or artesian wells. Even the
rocks and soils are permeated with water, while
plants and animals are largely constituted of it;
five sixths of a living human body is simply water.
How like a world prepared for a great conflagration our earth is, when we remember that eight
ninths of all this water, massed on its surface,
running as rivers or accumulated as lakes within
its crust, permeating its soils and rocks, constituting largely its vegetation and animals, is the
fearful furnace, oxygen gas! And yet further,
hydrogen, constituting the other ninth part of
water, is itself also combustible, producing an
intense heat, and when united with a proper proportion of air or oxygen, its combustion is in-
stantaneous and explosive. The Bible speaks of a
"loud noise" at the final conflagration. Another
striking fact is to be noted: the intensest heat by
far yet evolved by the blowpipe is by the combustion of those two gases, oxygen and hydrogen,
constituting water, so everywhere pervasive in the
earth.

And what may well astonish us, Dr. Robert Hare, of Philadelphia, the inventor of the oxy-hydrogen blowpipe, observed that "the most intense heat attainable was generated when the proportions of the gases were the same as in water."

Yet further, the solid crust of our globe — the earths and rocks themselves — holds in combination vast masses of this oxygen fuel. It is estimated that nearly one half of our globe is oxygen. For example, a single ounce of the peroxide of manganese contains one hundred and twenty-eight cubic inches of oxygen gas; every half-ounce of chlorate of potash affords two hundred and seventy cubic inches of pure gas. What an immense volume of this furnace gas, from even the earths and the rocks beneath us, is ready to burst forth at the touch of the competent finger!

Again, it is estimated that, descending from the surface of the earth towards the centre, a rise of 1° Fahrenheit occurs for every fifty-eight feet. "If, then," says Dr. Draper, "the increase of heat is only 100° per mile, at a depth of ten miles everything must be red-hot, and at thirty or forty miles below the surface in a melted state." This may startle us: our globe to-day a lava ball 8,000 miles in diameter, except a little outside fringe, greatly thinner in proportion to the inner bulk than shell of egg to inner bulk! The smoke and cinders and steam and glowing lava thrust out from an Ætna, a Vesuvius, a Mauna Loa, are but little warning whispers to the dweller on the earth — the insect, man — of the wild, heaving fire-ocean beneath his feet; while the continuous earthquakes and the rending of the earth in fissures, preceding the volcanic outflow, declare to man the frailty of the bridge which tremulously sustains him over the abyss of fire. Or regard, as some, the earth's crust 2,500 miles in thickness; in the centre a solid nucleus; between crust and nucleus, fire oceans, — even then our standing place, in its shaking, swaying, fracturing, seems frail.

When we thus in the light of science look in upon the fire-ocean beating wildly beneath our feet, when we consider the frailty of the bridge on which we stand, when we call to mind the combustible and explosive elements with which we are encompassed and with which our frail bridge is honeycombed, — rather than be astonished that the Bible says the world is " reserved for fire," we may, with the ancient naturalist Pliny, and the modern scientist Lyell, wonder that a frame of things so like a powder magazine stands for a moment unexploded.

Mr. R. A. Proctor, in his latest volume, "The Flowers of the Sky," speaks of another source whence may come by fire, destruction to all terrestrial life. Each star is a sun, in general respects similar to ours, each glowing with intense heat. Mr. Proctor notes that in 1866, a star in the constellation Northern Cross, suddenly shone with eight hundred times its former lustre, afterward rapidly diminishing in lustre : and that in 1876 a new star in the constellation Cygnus became visible, subsequently fading again so as to be only perceptible by means of a telescope ; the lustre of this star increased five hundred to many thousand times (according to data assumed). Proctor claims that should our star, the sun, similarly increase in lustre only one hundred times, the glowing heat would destroy all vegetable and animal life on our earth, even "the stubborn animalcules and the lowest forms of vegetation." But what a fearful furnace would our globe be in, if the sun's lustre increased "many thousand times," as possibly did that of the star in Cygnus!

As these star worlds of the Northern Cross and Cygnus suddenly shone out with unwonted lustre to the observer on the far-distant earth, so there approaches a day, a day which shall come all suddenly, a day perhaps near, when our globe to the distant observers on other spheres shall all suddenly shine forth with unwonted lustre; the moment then come when "the elements shall melt with fervent heat, the earth and the works that are therein be burnt up, and the heavens pass away with a great noise."

Says Principal Dawson ("Princeton Review," November, 1879), "It is a question raised by certain expressions of Scripture, whether the world (morally) will again fall into the condition in which it was before the Flood. 'As it was in the days of Noah,' we are told, 'so shall it be when the Son of Man comes to judgment.' To bring

the world into such a state, it would require that
it should shake off the superstitions, fears, and
religious hopes which now affect it; that it should
practically cast aside all belief in God, in morality, and in the spiritual nature and higher des-
tiny of man; that it should devote itself entirely
to the things that belong to the present life, and
the pursuit of those should be influenced by nothing higher than a selfish expediency. Then
would the earth again be filled with violence, and
again would it cry unto God for punishment, and
again would he say that his 'spirit should no
longer strive with man,' and that 'it repented him
that he had made man upon the earth.' Who
shall say that this is impossible? On the contrary, do we not see in the materialistic philosophy, in the cold, calculating policy, the profound
selfishness, and the proud self-confidence of the
more civilized races in our times, [in widespread
nihilism, which shouts 'Down with all government, down with the family, down with all
morality, down with God!'] indications of the
same spirit which was in the antediluvians?
Should it come to pass that this spirit should

again prevail, it might happen that God, who has so much patience with the follies, the superstitions, and the baser appetites of humanity, might again direct his judgments against this higher and more stupendous form of iniquity; and as the earth that then was perished by *water*, so that which is now might, in consideration of the clearer light it has abused and the greater privileges it has despised, be visited with *fire*, reserved against 'the day of judgment and perdition of ungodly men,' and *which nature can in many ways provide.*"

(*b.*) Let it be that the earth is melted by fire, and all things therein burnt up: what becomes of the burnt material, since no atom of matter is annihilated?

To-day this globe wheels through space wrapped in sin as in a garment; the vile poison of its filthy moral envelope permeates in God's eye the material earth itself; all has become defiled. This is the devil's work, through his seduction to sin of our first parents. But "the Son of God was manifested that he might destroy the works of the devil." Christ's work in this world is for the

devil's work an antidote, will be a perfect antidote
except where resisted by the will of the moral
creature. Christ's work, then, must include the
purification and restitution of the material earth.
Christ's purification of the earth by fire, and his
restitution of it, does not mean simply his putting
of it into a seven times heated furnace for a little
time, and then leaving it forever a blackened
desolation in God's universe. Restitution of the
earth by Christ means something infinitely better
than this; means the bringing of it out from under
curse a pure thing, reconstructing it into a new
Eden, making it once again the happy home of
holy beings. Says Isaiah, "Behold, I create new
heavens and a new earth." Peter tells us this
promise in his time was yet unfulfilled, and says
expectantly, "We, according to his promise, look
for new heavens and a new earth wherein dwelleth
righteousness" (dwell the righteous). So, later,
the Revelator, still looking forward, says, "I saw
a new heaven and a new earth, for the first heaven
and the first earth were passed away, and I, John,
saw the holy city, New Jerusalem, coming down
from God out of heaven, prepared as a bride

adorned for her husband"; and he tells us this
New Jerusalem, coming down out of heaven to the
new earth, is the "bride, the Lamb's wife," *i. e.*,
entire redeemed Israel. The restored new earth
is here declared to be the future home of the
redeemed; part of their glad, prophetic song to-
day in heaven is (Rev. v. 10), "We shall *reign
on the earth.*"

To this consummation the promise of God to
the Jews of the inheritance of Canaan points, as
also analogically does the past history of the
earth.

When we keep in mind that the Israelites were
a typical people, this renders their history intensely significant, interesting. Promises made
and fulfilled to them were not in that first fulfilment exhausted of all they contained; this first
fulfilment was simply typical; a mere shadow of
the substance in the promise; a narrow, small,
meagre thing, compared with the true, full, large
God-thought in its ultimate treasure, in its fulfilment to antitypical Israel. All this holds true
of the promise made to Abraham and his seed of
possession of Canaan. That promise to typical

Israel of a little bit of land, rich in the finest of the wheat, flowing with milk and honey, speaks to the antitypical Israel of an inheritance wider than ancient Canaan's narrow limits, of a land richer and of better fruits; is promise of God to his children universal, that one day they shall possess the earth in its entire compass, in all its riches, and in its every fountain of pleasure.

The Bible clearly sustains this enlarged view of the old promise of Canaan to Abraham and his seed. The Bible declares that the promise was never fulfilled in all its significance to them. Says Stephen, God gave Abraham " none inheritance in it, no not so much as to set his foot on; yet he promised that he would give it to him for a possession." Says Hebrews, " By faith Abraham sojourned in the land of promise, as in a strange country, dwelling in tabernacles with Isaac and Jacob, the heirs with him of the same promise. These all died in faith, not having received the promises but having seen them afar off, and confessed that they were strangers and pilgrims on the earth." And after naming a host of other worthies, the chapter thus ends: " These all, having obtained a good report through faith, received not the promise : God having provided some better thing for us, that they without us should not be made perfect"; *i. e.*, these worthies never entered into possession of the promise of Canaan in its deepest significance, never will until the entire company of God's redeemed, antitypical Israel shall enter upon the possession of the new earth, antitypical Canaan, — led into it under Jesus, the antitypical Joshua, — reigned over by God's anointed, Christ, the antitypical David.  Then is the time when the "stone" which Daniel saw "cut out of the mountain without hands," having dashed in pieces every opposing kingdom, shall "fill the whole earth"; then is the time, and not till then, when "the earth shall be full of the knowledge of the Lord, as the waters cover the sea"; then is the time when Jesus to his people shall fulfil his own words, "The meek shall inherit the earth."

This idea of the earth's purification in the future — renewal, being lifted into a higher condition, condition fitted for a higher order of beings, peopled with beings of a higher order — is sustained analogically by the past physical history of the earth.

In the past we see the earth in a state of fusion,
azoic. Then the chaotic water waste, azoic;
"light," not "lights" yet. "This air which we
breathe so freely was in the beginning so loaded
with poisonous carbonic acid as to be unfit for the
sustenance of organisms. At that time, therefore,
the earth was uninhabited. Gradually, by a subtile chemical process, immense quantities of the
carbonic acid was withdrawn and united with lime
to form immense beds of limestone, which still
form much of the strata of the earth. The atmosphere was now suitable for the growth of vegetation and the life of lower water-breathing animals,
but not for the higher air-breathing animals.
Therefore, when all the other preparatory arrangements were made for the introduction of
these higher animals, immense quantities of carbonic acid were again silently withdrawn from the
air by the luxuriant vegetation of the coal period,
— the carbon forming coal, the oxygen returning
to the air. Again during the Secondary period,
and again during the Tertiary period, the same
process was repeated; carbonic acid withdrawn
from the air, the carbon laid up as coal, and the

oxygen returned to the air." This change and rise of character in the atmosphere, progressing through untold centuries till the appearance of man, is type of changes to higher forms which have taken place in other physical conditions of the earth. And these upward changes in the physical conditions of the earth were ever followed by the appearance of new and higher grades of creatures upon the earth; the climax of physical conditions reached, the climax of the creature world appeared, — man.

At this point the earth was given into the hand of man "to dress and keep," and develop quietly into God's ideal. But man rebelled against the divine plan, refused "to dress and keep it," — sinned. Said God, " Cursed is the ground for thy sake." This curse must in some way have affected the earth for evil, thrust it backward from its possible development under man's " dressing and keeping," — perhaps paralyzed in it that progressively developing force which, acted on by man in obedient co-operation with God, would have caused it to pass on quietly into a very much higher condition than its present, fitted

it for the joyous habitation of man in the high
condition of a nature and a holiness developed
uninterruptedly from the germ of his primal in-
nocence and God-image.  Adam utterly failed to
accomplish this God-intrusted work, — "he fell,
and dragged the earth down with him into devas-
tating ruin.  He thus became absolutely incapable
of fulfilling his mission.  Hence Christ, the second Adam, came in the stead of man, to renew
and complete what man had destroyed and failed
of accomplishing; to lead the world on to its
God-ideal development and intended place in the
universe.  But this can now no longer be done in
the method originally intended, by quiet, gradual
organic development; this method has been disturbed and forever rendered impossible by the
man wrapping himself and the earth in the garment of sin, and drawing in upon the " ground "
God's paralyzing curse.  A new method of development and perfecting of the material earth and
its creatures must now be adopted by the new
Adam, delegated of God with ample powers for
his work.  This now can only be accomplished
by the breaking out, in the appalling catastrophe

of a burning, melting, furnace world, of the con-
suming and purifying fires of the last day; and
from out those flaming elements, purified from all
dross and defilement, shall issue a new earth, in
the perfection of God's ideal, — a perfect earth,
fitting abode for man perfect in Christ, — God's
ideal in the material earth and God's ideal in its
creature inhabitant at last reached." Reached,
but reached after, oh, how long a time since the
first atom of the primitive nebula, whence came
our solar system, was laid! A little gleam of
light is given here on the significance of that
word of Peter, "A thousand years is with the
Lord as one day," and on the significance of that
other word, "Eternity"!

"In the history of the earth which we inhabit,
mollusks, fishes, reptiles, mammals, had each in
succession their periods of vast duration; and then
the human period began, — the period of a fellow-
worker with God, created in God's own image.
What is to be the next advance? Is there to be
merely a repetition of the past, an introduction
a second time of man made in the image of God?
No. The geologist, in those tables of stone which

form his records, finds no example of dynasties once passed away, again returning. There has been no repetition of the dynasty of the fish, of the reptile, of the mammal. The dynasty of the future is to have glorified man for its inhabitant; but it is to be the dynasty of the ' *Kingdom*,' — not of glorified man made in the image of God, but of God himself in the form of man. In the doctrine of the two conjoined natures, human and divine, and in the further doctrine that the terminal dynasty is to be peculiarly the dynasty of HIM in whom the natures are united, we have that required progression beyond which progress cannot go. We find the point of elevation never to be exceeded meetly coincident with the final period never to be terminated; the infinite in height harmoniously associated with the eternal in duration. Creation and Creator meet at one point and in one person. The long ascending line from dead matter to man has been a progress Godwards; not an asymptotical progress, but destined from the beginning to furnish a point of union: and occupying that point as true God and true man, as Creator and created, we recognize the adorable Monarch of all the future, — the God-man, the antitypical David. ("Testimony of the Rocks," 178.)

"Formerly, when Jehovah laid the foundations of the earth, the morning stars, beholding with adoring wonder, sang together in choral songs of praise; and as the Eternal Word, full of grace and truth, left the throne of glory to clothe himself in flesh and blood, then swelled in higher and fuller notes the chorus of the heavenly host, 'Glory to God in the highest, and on earth peace, good-will to men!' In the future, also, when the son of man shall come again in the clouds, surrounded by all the glory of his eternal Godhead, to renew the heavens and the earth, consummate all things, and take possession of the Everlasting Kingdom, then shall those sacred messengers of his might and goodness, whose bosoms are thrilled with unspeakable joy at every new token of the spread of God's kingdom upon the earth, behold with adoring wonder the full development of those heaven-born mysteries they now desire to look into, and in purer tones and loftier chorus shall they sing their eternal hallelujahs,"— as they see their Lord triumphant over all his enemies, the devil's work on the earth destroyed, the earth restored to its pristine purity, beauty, full powers, the marriage supper of the Lamb come, and the Bridegroom and the bride enter into possession of the eternal inheritance, — the entire house of Israel possessing in its fulness the promise of Canaan at last. And this is the only "return" of the Jews; this is the restoration of Israel.

This very earth we now live in is thus to be God's redeemed people's eternal home. That very world in which Christ was crucified is the world in which Christ shall reign. That very world in which Christ's people were counted the "offscouring of all things," scourged, imprisoned, cast to wild beasts, drowned, burned, beheaded. racked to death, is the very world which shall one day be all their own, and in which they shall live and reign as priests and nobles and kings in a kingdom eternal. That very world in which the Christian has been once the slave of sin, has struggled with corruption, has endured the pains of sickness, has wept his tears beside the still forms of his loved ones, has himself gone down

in the conflict with the last enemy, is the very
world on whose renewed and purified soil he shall
walk a holy being, pure and free as an angel of
God. He shall nevermore say, " I am sick ! " the
pierced hand shall wipe away all tears from his
eyes; neither shall he die any more, but trium-
phantly shall sing, "O death, where is thy
sting? O grave, where is thy victory?" Rest,
sweet rest, now, to the child of the Great King,
beautiful robes, abundance at his Father's table,
large possessions, a splendid mansion, home now,
home forever !

(c.) What kind of a world shall the renewed
earth be?

When God made man and placed him on this
earth, the man and his dwelling-place were
adapted to each other. When man is simply
made a perfect man in Christ, the *general aspect*
of the earth need not be greatly changed. Had
primitive man obeyed God's words, " Dress it, keep
it," refrained from sin, and no curse come in upon
the " ground," blotting out from it some intrinsic
power of goodness, man might perhaps have led
it on quietly to a higher development, a com-

plete transformation, transfiguration; left now, this, for the second Adam to accomplish in the great conflagration and restoration. In the past, our world has ever been passing from lower to higher phases physically, preparatory to its becoming the abode of ever more highly developed inhabitants. When it comes out from its melting in fire, in its reconstruction the earth may be, in its material adaptations, furnishing, scenery, elevated in every way; be fitted, as never in all its past history, for the new and higher life now to enter it, — man in his spiritual body, a transformation and a transfiguration passed upon the earth and all its furnishing in its baptism of fire and in its renewal, similar to that which the matter of the bodies of the saints living on the earth at the sounding of the last trump shall undergo in the twinkling of an eye, and the " changed " earth and furnishing be perfectly, beautifully, thrillingly adapted to the saints in their bodies " changed."

But while we conceive of the future earth as elevated, passed on to a higher condition, we are not to conceive of it as divested of the qualities of materialism; it shall be still a genuine, solid earth.

Says Dr. Chalmers, "The object of the adminis-
tration we sit under is to extirpate sin, not to
sweep away materialism. By the convulsions of
the last day materialism may be shaken and broken
down from its present arrangements, and thrown
into such fitful agitations as that the whole of its
existing framework shall fall to pieces, and by a
heat so fervent as to melt its most solid elements
may be utterly dissolved ; and thus may the earth
again become without form and void, but without
one particle of its substance going into annihila-
tion. Out of the ruins of this second chaos shall
another heaven and another earth be made to
arise, and a new materialism, with perhaps other
aspects of magnificence and beauty, emerge from
the wreck of this mighty transformation ; and the
world be peopled as before, with the varieties of
material loveliness, and space be again lighted up
into a firmament of material splendor."

To him who shall enter into its possession, the
new earth in its adaptations, its furnishing, its
beauty, its employments, its varied, intense, ever-
enlarging river of pleasures, shall be a gift worthy
the universal Sovereign to his child beloved.

Says Saurin, "Could I extract the choicest digni-
ties and fortunes, could I inhabit the most temperate clime and the most pleasant country,
could I choose the most benevolent hearts and
the wisest minds, could I take the most happy
temper and the most sublime genius, could I
cultivate the sciences and make the fine arts
flourish, could I collect and unite all that could
please the passions, and banish all that could give
pain, — a life formed on this plan, how likely to
please us! How is it that God, who has resolved
to render us one day happy, does not allow us to
continue in this world, and content himself with
uniting all these happy circumstances in our
favor? 'It is good to be here.' Oh, that he
would allow us here to build our *tabernacles!*
Ah, a life formed on this plan might indeed
answer the ideas of happiness which feeble and
finite geniuses form, but such a plan cannot even
approach the designs of an infinite God. No, all
the charms of this society, of this fortune, and of
this life; no, all the softness of these climates,
and of these countries; no, all the benevolences
of these hearts, and all the friendship of these

minds; no, all the happiness of this temper, and all the sublimity of this genius; no, all the secrets of the sciences, and all the discoveries of the fine arts, — all the attractions of these societies and all the pleasures of the passions have nothing, I do not say which *exhausts* the love of God and its treasure-thoughts for his child, I do not say which *answers*, I venture to say which *approaches* it. To accomplish this love, to lead the redeemed into the fulness of its treasure-thought, there must be another world; there must be new heavens and a new earth; there must be objects far more grand."

> " Where the faded flowers shall freshen,
>     Freshen never more to fade;
> Where the shaded sky shall brighten,
>     Brighten never more to shade;
> Where the sun-blaze never scorches,
>     Where the star-beams cease to chill,
> Where no tempest stirs the echoes
>     Of the wood or wave or hill. . . .
> Where no shadow shall bewilder;
>     Where life's vain parade is o'er;
> Where the sleep of sin is broken,
>     And the dreamer dreams no more.
> Where the bond is never severed;
>     Partings, claspings, sob and moan,
> Midnight working, twilight weeping,
>     Heavy noontide, — all are done.

Where the child has found its mother,
  Where the mother finds the child;
Where dear families are gathered,
  That were scattered on the wild. . . .
Where the hidden wound is healèd;
  Where the blighted life reblooms;
Where the smitten heart the freshness
  Of its buoyant youth resumes. . . .
Where we find the joy of loving
  As we never loved before,
Loving on, unchilled, unhidden,
  Loving once, forevermore."

www.ingramcontent.com/pod-product-compliance
Lightning Source LLC
Chambersburg PA
CBHW021816190326
41518CB00007B/616